METHODS IN MOLECULA

CW01559167

Series Editor
John M. Walker
School of Life Sciences
University of Hertfordshire
Hatfield, Hertfordshire, AL10 9AB, UK

For further volumes:
http://www.springer.com/series/7651

Recombinant Protein Production in Yeast

Methods and Protocols

Edited by

Roslyn M. Bill

School of Life & Health Sciences and Aston Research Centre for Healthy Ageing
Aston University, Birmingham, UK

 Humana Press

Editor
Roslyn M. Bill
School of Life & Health Sciences
and Aston Research Centre for Healthy Ageing
Aston University
Birmingham, UK

ISSN 1064-3745   e-ISSN 1940-6029
ISBN 978-1-4939-5871-9   ISBN 978-1-61779-770-5 (eBook)
DOI 10.1007/978-1-61779-770-5
Springer New York Dordrecht Heidelberg London

Humana Press is part of Springer Science+Business Media (www.springer.com)

Dedication

For Martin and Alice

Preface

Advances in our understanding of the biology of the cell rely on our increasing knowledge of protein behaviour and the complex interplay of proteins with other biomolecules. Proteins are essential components of living organisms and have a role in virtually every cellular process: they are enzymes, form cellular scaffolds, and are central to signalling, transport, and regulatory functions. To study these diverse roles, it is necessary to be able to work with sufficient quantities of suitably stable and functional protein samples. While some proteins can be isolated from native sources for this purpose, many cannot as they are either intrinsically unstable or are present in impractically low quantities. Moreover, the study of mutant forms of a given protein is often central to understanding its structure and activity, and such mutants must be synthesized in vitro.

Since the 1970s, recombinant DNA technologies have provided a solution to producing proteins in non-native host cells. However, the attainment of high yields of recombinant proteins, particularly recombinant membrane proteins, is an enduring bottleneck in the post-genomic sciences that is only now being addressed in a truly rational manner. Several host systems have been developed for the production of recombinant proteins, ranging from prokaryotes, such as *Escherichia coli*, to higher eukaryotes, such as mammalian cell-lines. This book describes strategies and protocols for the use of yeast as a production host. With its extensive literature and sequenced genome, the well-known yeast species, *Saccharomyces cerevisiae*, provides many opportunities to optimize the functional yields of a target protein of interest. *Pichia pastoris*, for which there is now a freely available genome sequence, is a popular alternative. Both have been used for the industrial production of pharmaceutical proteins and are described here.

In the last few years, significant advances have been made in understanding how a yeast cell responds to the stress of producing a recombinant protein, and how this information can be used to engineer improved host strains. The molecular biology of the expression vector, through the choice of promoter, tag, and codon optimisation of the target gene, is also a key determinant of a high-yielding protein production experiment. The use of statistical approaches to examine how parameters such as medium composition, growth variables (e.g., temperature, oxygen availability), and the precise details of the induction regime influence recombinant protein yield has also been implemented more widely by researchers in the field.

This book examines the process of preparation of expression vectors, transformation to generate high-yielding clones, optimisation of experimental conditions to maximise yields, scale-up to bioreactor formats, and disruption of yeast cells to enable the isolation of the recombinant protein prior to purification. I hope the chapters describing these steps enable you to adopt yeast as a protein production host for your research.

Birmingham, UK   *Roslyn M. Bill*

Contents

Contributors

FATIMA ALKHALFIOUI • *Département Récepteurs et Protéines Membranaires, Centre National de la Recherche Scientifique, Ecole Supérieure de Biotechnologie de Strasbourg, Illkirch, France*

ZHARAIN BAWA • *School of Life & Health Sciences and Aston Research Centre for Healthy Ageing, Aston University, Birmingham, UK*

ROSLYN M. BILL • *School of Life & Health Sciences and Aston Research Centre for Healthy Ageing, Aston University, Birmingham, UK*

NICKLAS BONANDER • *Department of Chemical and Biological Engineering, Chalmers University of Technology, Gothenburg, Sweden*

NAGAMANI BORA • *School of Life & Health Sciences and Aston Research Centre for Healthy Ageing, Aston University, Birmingham, UK*

BERNADETTE BYRNE • *Division of Molecular Biosciences, Imperial College London, London, UK*

STEPHANIE P. CARTWRIGHT • *School of Life & Health Sciences and Aston Research Centre for Healthy Ageing, Aston University, Birmingham, UK*

MICHELLE CLARE • *School of Life & Health Sciences and Aston Research Centre for Healthy Ageing, Aston University, Birmingham, UK*

RICHARD A.J. DARBY • *Nuffield Department of Clinical Laboratory Sciences, John Radcliffe Hospital, University of Oxford, Oxford, UK*

MARVIN V. DILWORTH • *School of Life & Health Sciences and Aston Research Centre for Healthy Ageing, Aston University, Birmingham, UK*

DAVID DREW • *Division of Molecular Biosciences, Membrane Protein Crystallography Group, Imperial College, London, UK*

CEDRIC FIEZ-VANDAL • *Département Récepteurs et Protéines Membranaires, Centre National de la Recherche Scientifique, Ecole Supérieure de Biotechnologie de Strasbourg, Illkirch, France*

ADRIEN GRAS • *Division of Molecular Biosciences, Imperial College London, London, UK*

KRISTINA HEDFALK • *Department of Chemistry and Molecular Biology, University of Gothenburg, Göteborg, Sweden*

MOHAMMED JAMSHAD • *School of Biosciences, University of Birmingham, Birmingham, West Midlands, UK*

HYUN KIM • *Laboratory of Membrane Biology, School of Biological Sciences, Seoul National University, Seoul, South Korea*

EDMUND R.S. KUNJI • *The Medical Research Council, Mitochondrial Biology Unit, Cambridge, UK*

CHRISTEL LOGEZ • *Département Récepteurs et Protéines Membranaires, Centre National de la Recherche Scientifique, Ecole Supérieure de Biotechnologie de Strasbourg, Illkirch, France*

MAGDALENA MARTINEZ • *Division of Molecular Biosciences, Imperial College London, London, UK*

SHANE M. PALMER • *The Medical Research Council, Mitochondrial Biology Unit, Cambridge, UK*

SARAH J. ROUTLEDGE • *School of Life and Health Sciences, Aston University, Birmingham, UK*

SHWETA SINGH • *Division of Molecular Biosciences, Imperial College London, London, UK*

RENAUD WAGNER • *Département Récepteurs et Protéines Membranaires, Centre National de la Recherche Scientifique, Ecole Supérieure de Biotechnologie de Strasbourg, Illkirch, France*

MARTIN D.B. WILKS • *Smallpeice Enterprises Ltd, Warwickshire, UK*

Chapter 1

Optimising Yeast as a Host for Recombinant Protein Production (Review)

Nicklas Bonander and Roslyn M. Bill

Abstract

Having access to suitably stable, functional recombinant protein samples underpins diverse academic and industrial research efforts to understand the workings of the cell in health and disease. Synthesising a protein in recombinant host cells typically allows the isolation of the pure protein in quantities much higher than those found in the protein's native source. Yeast is a popular host as it is a eukaryote with similar synthetic machinery to the native human source cells of many proteins of interest, while also being quick, easy, and cheap to grow and process. Even in these cells the production of some proteins can be plagued by low functional yields. We have identified molecular mechanisms and culture parameters underpinning high yields and have consolidated our findings to engineer improved yeast cell factories. In this chapter, we provide an overview of the opportunities available to improve yeast as a host system for recombinant protein production.

Key words: Recombinant protein production, Yeast, Strain engineering, Bioprocess control

1. Introduction

The development of recombinant protein production hosts that can be used to produce a wide range of targets is a key area of research (1). This is particularly true for the production of membrane proteins, which are high value targets in the drug discovery pipeline, and which cannot yet be produced in high yields in a predictable manner (2, 3). Yeast species, especially *Pichia pastoris* and *Saccharomyces cerevisiae* (2, 4–6) have already been identified as one of the most important components of a matrix of protein production hosts (7), and have contributed a substantial number of functional (8) recombinant eukaryotic membrane proteins in very high yields (e.g. ref. 9) thus enabling high resolution structure determination (10–13).

Roslyn M. Bill (ed.), *Recombinant Protein Production in Yeast: Methods and Protocols*, Methods in Molecular Biology, vol. 866, DOI 10.1007/978-1-61779-770-5_1, © Springer Science+Business Media, LLC 2012

Yeasts are well-characterised organisms and can be cultured very cheaply and easily in large quantities. They are also straightforward to manipulate genetically: publicly available sequences are available for both *P. pastoris* (14) and *S. cerevisiae* (15). As yeasts are eukaryotes, they have protein-processing and post-translational modification mechanisms related to those found in mammalian cells. However, they contain ergosterol rather than cholesterol in their membranes and recombinant proteins produced using yeast may be decorated with high-mannose type sugars, which are not native to mammalian cells (2). Fortunately, these problems have been overcome, in part, through strain engineering (16, 17). Despite these potential limitations, recombinant pharmaceuticals, including insulin (18) and several vaccines (e.g. ref. 19) have all been produced in yeasts for commercial use, demonstrating their importance to the pharmaceutical industry. In order to optimise their productivity, two complementary approaches have been taken: engineering the yeast strain and modifying the culture or "bioprocess" conditions. Both are discussed in this chapter and are exemplified by studies on *S. cerevisiae*.

2. Host Cell Engineering to Increase Recombinant Protein Yields

Strategies to increase the recombinant protein productivity of host cells have been examined in both prokaryotic and eukaryotic cells. For example, a strain of *Escherichia coli* with high cardiolipin content (14% compared with 3% in wild-type strains) has been found to produce increased yields of some proteins (20, 21). *Lactococcus lactis* has also been evolved to generate strains with modest improvements in their yield properties (22). In both cases, mutations in the promoter system, and not in any in vivo cellular components of the translation or folding machinery, were found to have slowed the protein synthesis rate, presumably allowing the newly synthesised protein to be accommodated into the endoplasmic reticulum with a minimum of inclusion body formation and cell death.

We and others have used a more systematic approach (23–25) to optimise a host cell for recombinant protein production, with the specific goal of understanding the molecular barriers to achieving high yields. We therefore looked at the global host cell response to producing a membrane protein in *S. cerevisiae* (23). Using transcriptome arrays, we examined two different growth conditions that both led to relatively low protein yields (compared with normal conditions) and looked for changes in mRNAs that occurred in the same direction in both sets. This highlighted a set of genes, which was validated against those identified when comparing the single high-yielding growth condition with normal conditions. Genes that were down-regulated under low-yielding conditions were up-regulated under high-yielding conditions and vice versa (26). This allowed us to identify genes that influenced the yield per cell (27): three are known components of the transcriptional SAGA (*GCN5* and *SPT3*) and mediator (*SRB5*) complexes, while a fourth, *BMS1*, is involved in ribosome biogenesis (26). Interestingly, there was an increase in *BMS1* transcript number compared to wild-type in all our high-yielding host strains. In particular, we were able to tune the expression of *BMS1* to maximise yields (26).

Work such as this has highlighted other relevant pathways for further study. For example in producing a recombinant membrane protein, there may be a benefit in altering the membrane lipid content in the host cell (25), as has been seen previously for *E. coli* strains (20, 21). Opi1 protein acts as a transcriptional suppressor, affecting the synthesis of phospholipids (28). An *opi1Δ* strain therefore contains increased phospholipid concentrations compared with wild-type strains (29). A study of strains with altered lipid composition in our own laboratory (e.g. *opi1Δ*, *bio2Δ*, and *erg6Δ*) revealed that *opi1Δ* yielded eight times more protein than the corresponding wild-type strain (unpublished). Similar studies in *P. pastoris* on the global response to producing a recombinant protein (e.g. ref. 30) should soon lead to similar advances.

3. Optimising Culture Conditions to Increase Recombinant Protein Yields

Although the potential returns are high, engineering a host organism to increase recombinant protein yields can be expensive and time-consuming, especially if strategies are based on transcriptome analysis. An alternative, complementary, and potentially cheaper route is to optimise the experimental culture conditions, ideally using a statistical design of experiments approach (see Chapter 11). The development of growth media for mammalian cell culture has led the way in this area. For example, medium composition has been key to maintaining viable, high density cultures to obtain increased monoclonal antibody titres (31) or to be able to use chemically defined serum-free medium (32, 33).

Variations in the composition of the culture medium, its pH, its temperature, the availability of dissolved oxygen and the details of the induction regime (e.g. concentration of inducer, as well as the point and duration of induction) may all lead to improvements in the productivity of yeast cultures. For example, the yield of active G protein-coupled receptors (GPCRs) produced in *P. pastoris* was increased in over half the cases when the growth medium was supplemented with 2.5% dimethyl sulfoxide (DMSO; 16 out of 20) or 0.04 mg/mL L-histidine (12 out of 20) (34). However, in another study, the human tetraspanin protein, CD81, was produced and purified from *P. pastoris* (35), but supplementation with 2.5%

DMSO did not increase yields (unpublished). Often, however, the induction regime in *P. pastoris* has a much greater effect on yield than the growth conditions (36).

For cultivating *S. cerevisiae*, the most commonly used media formulations in our laboratory are CBS (devised by the Centralbureau voor Schimmelcultures, the Netherlands; (37)) and those based on Yeast Nitrogen Base (YNB; (38)). YNB-based media are simpler to make than CBS, and originate from the work of Burkholder and Wickermann who used alterations in growth media to discriminate between yeast strains (39–41). For use with typical auxotrophic laboratory strains, both formulations require additional supplementation with an amino acid mixture, adenine and uracil (e.g. complete supplement mixture or CSM), as well as a carbon source (see Notes 1 and 2). Growth to medium cell densities ($OD_{600} < 50$) is possible in 2× CBS medium supplemented with 20 or 50 g/L glucose as the carbon source (42, 43), while growth in a YNB-based medium typically yields less biomass and thus less recombinant protein. The use of alternative carbon sources, such as ethanol, has also been a route to increased protein yields (44). Overall, CBS medium supports a higher growth rate than YNB-based medium, but is much more labour intensive to prepare (23).

It has become clear in our own studies that YNB-based media are not ideal for protein production in *S. cerevisiae* (25). In order to combine the simplicity of the YNB-based formulation (see Note 1) with the improved performance of CBS medium (see Note 2), we have undertaken further optimisation studies. For example, since the membranes of the endoplasmic reticulum have a high content of phosphoinositol lipids, we examined the effect of supplementing a YNB-based medium with myo-inositol to the levels found in CBS (Table 1) on recombinant membrane protein yield. Initial data suggesting that yields were improved (unpublished) are consistent with our previous observations for the *opi1Δ* strain, above, as well as literature reports suggesting that myo-inositol is essential for relieving some cellular stresses caused by recombinant membrane protein production (45). Furthermore, supplementation with myo-inositol (as well as biotin; Table 1) was found to increase growth rates to values typical of CBS medium. Adding both myo-inositol and biotin also increased the growth rate, but less dramatically (Fig. 1). Addition of cobalt was found to decrease the growth rate even in the presence of myo-inositol and biotin, although in the presence of copper this decrease did not occur. Observations such as these could be followed up using a statistical design (46–48) to minimise the number of experiments required to examine all possible combinations of these medium components; especially, as metal ion requirements for *S. cerevisiae* are often strain specific (49) (see Chapter 11).

The beneficial effect of adding myo-inositol to YNB-based media on recombinant membrane protein yield may indicate that different medium components could be fine-tuned to relieve the stresses induced during recombinant soluble protein expression (50).

Table 1
Comparison of selected nutrient components in media based on CBS and YNB formulations

Nutrient	CBS (mg/L)	YNB (mg/L)	Factor difference (CBS:YNB)
Biotin	0.05	0.002	25
Myo-inositol	25	2	12.5
Cobalt chloride hexahydrate	0.3	0	∞
Zinc sulphate heptahydrate	4.5	0.4	11.25
Copper sulphate pentahydrate	0.3	0.062	7.5

S. cerevisiae requires a minimum concentration of Cu^{2+}, Zn^{2+}, and Fe^{2+} of 15, 200 and 150 µg/L, respectively (49). In YNB-based media, these concentrations are typically 16, 91, and 88 µg/L, respectively, and in CBS they are 76, 1,023, and 602 µg/L, respectively. Other metal ions not included in the YNB formulation are Co^{2+} and Ni^{2+}. Cobalt in particular is known to have an inhibitory effect on growth (52). Boric acid is present in YNB-based formulations at a concentration of 500 µg/L (8 µM) and is known to increase biomass yields during respiratory growth at 150 µM (53). The five nutrients with the largest difference in concentration between CBS and YNB media are listed

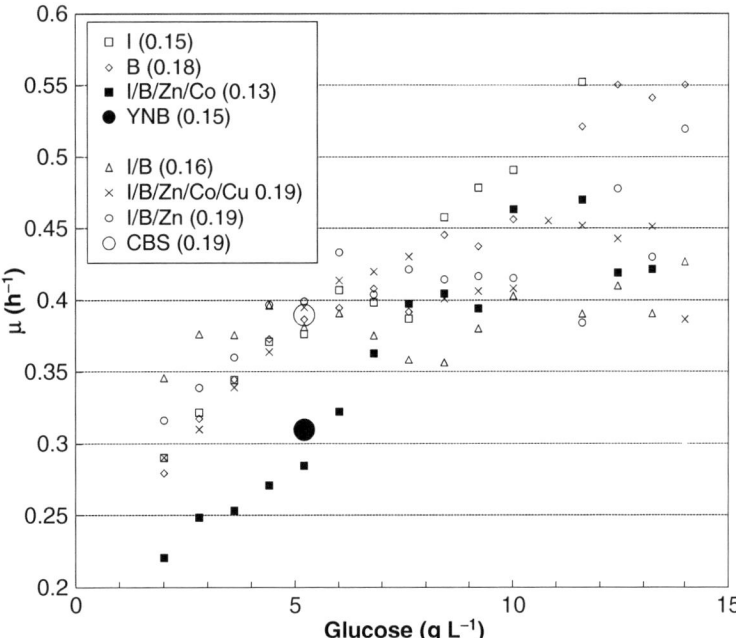

Fig. 1. Specific growth rates for *S. cerevisiae* strain BY4741 producing recombinant Fps1. The specific growth rate (μ; h^{-1}) was computed using the amount of base added in two adjacent 20 min segments in the 2–14 g/L residual glucose range (23). The average dry weight after consumption of the carbon source (at 55 h) was 0.17 g cell dry weight/g consumed glucose. This value increased by 0.04 g/g in the presence of myo-inositol and biotin. Media additions to YNB are as follows: myo-inositol (*open squares*); biotin (*open diamonds*); myo-inositol/biotin/ zinc/cobalt (*filled squares*); myo-inositol/biotin (*open triangles*); myo-inositol/biotin/zinc/cobalt/ copper (*crosses*); myo-inositol/biotin/zinc (*open circles*). YNB alone is shown as a *large filled circle*. 2× CBS is shown as a *large open circle*. Individual dry weight values (g/g) are listed in parenthesis after each condition ($n=2$; standard deviation < 0.008 for all data points).

For example, induction of the unfolded protein response (51) might require a different balance of nutrient components. Adding only the required nutrients at just the right concentration for maximum productivity should enable this to be achieved in the most cost- and time-effective manner.

4. Conclusions

YNB-based media are quick and easy to make, but do not support optimal recombinant protein yields. Supplementing the formulation with additional myo-inositol and biotin may overcome these limitations. This is especially true for recombinant membrane proteins and complements approaches based on strain engineering.

5. Notes

1. A YNB-based medium for a strain with a given auxotrophy is composed of 1.7 g/L YNB (without ammonium sulphate and amino acids, BIO101 #4027-012), 5 g/L ammonium sulphate and CSM minus the relevant component, as directed by the manufacturer (BIO101 #4510-333). Glucose is autoclaved separately as a 20% stock solution and added to a final concentration of 2%. If the glucose is autoclaved together with the ammonium sulphate and YNB, the solution will have a yellowish appearance. At a final $OD_{600} < 15$ this will not affect growth, but it is quite probable that nutrient limitation events will occur when growing to higher densities (using e.g. 5% glucose), due to glucose-induced reactions during autoclaving of the nutrients together. It is generally considered that the maximum glucose concentration that a YNB-based medium will support is 5%; at higher glucose concentrations, for example in a fed-batch culture, additional nutrients should be provided.

2. 2× CBS medium is composed of 10 g/L ammonium sulphate, 6 g/L potassium dihydrogen phosphate, 1 g/L magnesium sulphate heptahydrate plus twice the amount of CSM minus the relevant component directed (BIO101 #4510-333). Glucose is autoclaved separately as a 20% stock solution and added to a final concentration of 2%. Finally, 2 mL/L each of trace element solution and vitamin stock solution are added. 1 L trace element solution is composed of the following: 15 g EDTA, 4.5 g $ZnSO_4 \cdot 7 H_2O$, 1 g $MnCl_2 \cdot 4H_2O$, 0.3 g $CoCl_2 \cdot 6H_2O$, 0.3 g $CuSO_4 \cdot 5H_2O$, 0.4 g $Na_2MoO_4 \cdot 2H_2O$, 4.5 g $CaCl_2 \cdot 2H_2O$, 3 g $FeSO_4 \cdot 7H_2O$, 1 g H_3BO_3, and 0.1 g KI. The pH is maintained at 6.0 with 1 M NaOH throughout

the addition and finally adjusted to pH 4 with 1 M HCl prior to autoclave sterilisation and storage at 4°C in the dark. 1 L vitamin solution is composed of the following: 0.05 g biotin, 1 g calcium D-panthothenate, 1 g nicotinic acid, 25 g myo-inositol, 1 g thiamine hydrochloride, 1 g pyridoxol hydrochloride, and 0.2 g D-amino benzoic acid. The pH is maintained at 6.5 with 1 M HCl. The vitamin solution is filter-sterilised and stored as 20-mL aliquots at 4°C.

Acknowledgements

This work was supported by the European Commission via contract LSHG-CT-2004-504601 (E-MeP), LSHG-CT-2006-037793 (OptiCryst) and Grant 201924 (EDICT) to RMB. The BBSRC supports bioreactors and a flow microcalorimeter in the RMB Laboratory through an REI award.

References

1. Bill RM, Henderson PJ, Iwata S, Kunji ER, Michel H, Neutze R, Newstead S, Poolman B, Tate CG, Vogel H (2011) Overcoming barriers to membrane protein structure determination. Nat Biotechnol 29:335–340

2. Bill RM (2001) Yeast – a panacea for the structure-function analysis of membrane proteins? Curr Genet 40:157–171

3. Grisshammer R, Tate CG (1995) Overexpression of integral membrane proteins for structural studies. Q Rev Biophys 28:315–422

4. Gelperin DM, White MA, Wilkinson ML, Kon Y, Kung LA, Wise KJ, Lopez-Hoyo N, Jiang L, Piccirillo S, Yu H, Gerstein M, Dumont ME, Phizicky EM, Snyder M, Grayhack EJ (2005) Biochemical and genetic analysis of the yeast proteome with a movable ORF collection. Genes Dev 19:2816–2826

5. Reilander H, Weiss HM (1998) Production of G-protein-coupled receptors in yeast. Curr Opin Biotechnol 9:510–517

6. Sarramegna V, Talmont F, Demange P, Milon A (2003) Heterologous expression of G-protein-coupled receptors: comparison of expression systems fron the standpoint of large-scale production and purification. Cell Mol Life Sci 60:1529–1546

7. Freigassner M, Pichler H, Glieder A (2009) Tuning microbial hosts for membrane protein production. Microb Cell Fact 8:69

8. Karlgren S, Pettersson N, Nordlander B, Mathai JC, Brodsky JL, Zeidel ML, Bill RM, Hohmann S (2005) Conditional osmotic stress in yeast: a system to study transport through aquaglyceroporins and osmostress signaling. J Biol Chem 280:7186–7193

9. Nyblom M, Oberg F, Lindkvist-Petersson K, Hallgren K, Findlay H, Wikstrom J, Karlsson A, Hansson O, Booth PJ, Bill RM, Neutze R, Hedfalk K (2007) Exceptional overproduction of a functional human membrane protein. Protein Expr Purif 56:110–120

10. Jidenko M, Nielsen RC, Sorensen TL, Moller JV, le Maire M, Nissen P, Jaxel C (2005) Crystallization of a mammalian membrane protein overexpressed in Saccharomyces cerevisiae. Proc Natl Acad Sci USA 102:11687–11691

11. Long SB, Campbell EB, Mackinnon R (2005) Crystal structure of a mammalian voltage-dependent Shaker family K^+ channel. Science 309:897–903

12. Tornroth-Horsefield S, Wang Y, Hedfalk K, Johanson U, Karlsson M, Tajkhorshid E, Neutze R, Kjellbom P (2006) Structural mechanism of plant aquaporin gating. Nature 439:688–694

13. Horsefield R, Norden K, Fellert M, Backmark A, Tornroth-Horsefield S, Terwisscha van Scheltinga AC, Kvassman J, Kjellbom P, Johanson U, Neutze R (2008) High-resolution x-ray structure of human aquaporin 5. Proc Natl Acad Sci USA 105:13327–13332

14. De Schutter K, Lin YC, Tiels P, Van Hecke A, Glinka S, Weber-Lehmann J, Rouze P, Van de

Peer Y, Callewaert N (2009) Genome sequence of the recombinant protein production host *Pichia pastoris*. Nat Biotechnol 27:561–566

15. Goffeau A, Barrell BG, Bussey H, Davis RW, Dujon B, Feldmann H, Galibert F, Hoheisel JD, Jacq C, Johnston M, Louis EJ, Mewes HW, Murakami Y, Philippsen P, Tettelin H, Oliver SG (1996) Life with 6000 genes. Science 274:546, 563–567

16. Hamilton SR, Bobrowicz P, Bobrowicz B, Davidson RC, Li H, Mitchell T, Nett JH, Rausch S, Stadheim TA, Wischnewski H, Wildt S, Gerngross TU (2003) Production of complex human glycoproteins in yeast. Science 301:1244–1246

17. Hamilton SR, Gerngross TU (2007) Glycosylation engineering in yeast: the advent of fully humanized yeast. Curr Opin Biotechnol 18:387–392

18. Kjeldsen T, Ludvigsen S, Diers I, Balschmidt P, Sørensen AR, Kaarsholm NC (2002) Engineering-enhanced protein secretory expression in yeast with application to insulin. J Biol Chem 277:18245–18248

19. Siddiqui MAA, Perry CM (2006) Human papillomavirus quadrivalent (types 6, 11, 16, 18) recombinant vaccine (Gardasil (R)). Drugs 66:1263–1271

20. Miroux B, Walker JE (1996) Over-production of proteins in *Escherichia coli*: mutant hosts that allow synthesis of some membrane proteins and globular proteins at high levels. J Mol Biol 260:289–298

21. Arechaga I, Miroux B, Karrasch S, Huijbregts R, de Kruijff B, Runswick MJ, Walker JE (2000) Characterisation of new intracellular membranes in *Escherichia coli* accompanying large scale over-production of the b subunit of F(1)F(o) ATP synthase. FEBS Lett 482: 215–219

22. Linares DM, Geertsma ER, Poolman B (2010) Evolved *Lactococcus lactis* strains for enhanced expression of recombinant membrane proteins. J Mol Biol 401:45–55

23. Bonander N, Hedfalk K, Larsson C, Mostad P, Chang C, Gustafsson L, Bill RM (2005) Design of improved membrane protein production experiments: quantitation of the host response. Protein Sci 14:1729–1740

24. Griffith DA, Delipala C, Leadsham J, Jarvis SM, Oesterhelt D (2003) A novel yeast expression system for the overproduction of quality-controlled membrane proteins. FEBS Lett 553:45–50

25. Bonander N, Bill RM (2009) Relieving the first bottleneck in the drug discovery pipeline: using array technologies to rationalize membrane protein production. Expert Rev Proteomics 6:501–505

26. Bonander N, Darby RA, Grgic L, Bora N, Wen J, Brogna S, Poyner DR, O'Neill MA, Bill RM (2009) Altering the ribosomal subunit ratio in yeast maximizes recombinant protein yield. Microb Cell Fact 8:10

27. Bawa Z, Bland CE, Bonander N, Bora N, Cartwright SP, Clare M, Conner MT, Darby RA, Dilworth MV, Holmes WJ, Jamshad M, Routledge SJ, Gross SR, Bill RM (2011) Understanding the yeast host cell response to recombinant membrane protein production. Biochem Soc Trans 39:1113

28. Santiago TC, Mamoun CB (2003) Genome expression analysis in yeast reveals novel transcriptional regulation by inositol and choline and new regulatory functions for Opi1p, Ino2p and Ino4p. J Biol Chem 278:38723–38730

29. Jiranek V, Graves JA, Henry SA (1998) Pleiotropic effects of the opi1 regulatory mutation of yeast: its effects on growth and on phospholipid and inositol metabolism. Microbiology 144:2739–2748

30. Sohn SB, Graf AB, Kim TY, Gasser B, Maurer M, Ferrer P, Mattanovich D, Lee SY (2010) Genome-scale metabolic model of methylotrophic yeast *Pichia pastoris* and its use for in silico analysis of heterologous protein production. Biotechnol J 5:705–715

31. De Alwis DM, Dutton RL, Scharer J, Moo-Young M (2007) Statistical methods in media optimization for batch and fed-batch animal cell culture. Bioprocess Biosyst Eng 30:107–113

32. Stiens LR, Buntemeyer H, Lutkemeyer D, Lehmann J, Bergmann A, Weglohner W (2000) Development of serum-free bioreactor production of recombinant human thyroid stimulating hormone receptor. Biotechnol Prog 16:703–709

33. van der Valk J, Brunner D, De Smet K, Fex Svenningsen A, Honegger P, Knudsen LE, Lindl T, Noraberg J, Price A, Scarino ML, Gstraunthaler G (2010) Optimization of chemically defined cell culture media – replacing fetal bovine serum in mammalian *in vitro* methods. Toxicol In Vitro 24:1053–1063

34. Andre N, Cherouati N, Prual C, Steffan T, Zeder-Lutz G, Magnin T, Pattus F, Michel H, Wagner R, Reinhart C (2006) Enhancing functional production of G protein-coupled receptors in *Pichia pastoris* to levels required for structural studies via a single expression screen. Protein Sci 15:1115–1126

35. Bonander N, Jamshad M, Hu K, Farquhar MJ, Stamataki Z, Balfe P, McKeating JA, Bill RM (2011) Structural characterization of

CD81-Claudin-1 hepatitis C virus receptor complexes. Biochem Soc Trans 39:537–540

36. Holmes WJ, Darby RA, Wilks MD, Smith R, Bill RM (2009) Developing a scalable model of recombinant protein yield from *Pichia pastoris*: the influence of culture conditions, biomass and induction regime. Microb Cell Fact 8:35

37. Verduyn C, Postma E, Scheffers WA, Van Dijken JP (1992) Effect of benzoic acid on metabolic fluxes in yeasts: a continuous-culture study on the regulation of respiration and alcoholic fermentation. Yeast 8:501–517

38. Abelovska L, Bujdos M, Kubova J, Petrzselyova S, Nosek J, Tomaska L (2007) Comparison of element levels in minimal and complex yeast media. Can J Microbiol 53:533–535

39. Burkholder PR, McVeigh I, Moyer D (1944) Studies on some growth factors of yeasts. J Bacteriol 48:385–391

40. Wickerham LJ (1946) A critical evaluation of the nitrogen assimilation tests commonly used in the classification of yeasts. J Bacteriol 52:293–301

41. Wickerman LJ (1951) Taxonomy of yeasts. US Dept Agri Tech Bull 1029:1–56

42. Henricsson C, de Jesus Ferreira MC, Hedfalk K, Elbing K, Larsson C, Bill RM, Norbeck J, Hohmann S, Gustafsson L (2005) Engineering of a novel *Saccharomyces cerevisiae* wine strain with a respiratory phenotype at high external glucose concentrations. Appl Environ Microbiol 71:6185–6192

43. Ferndahl C, Bonander N, Logez C, Wagner R, Gustafsson L, Larsson C, Hedfalk K, Darby RA, Bill RM (2010) Increasing cell biomass in *Saccharomyces cerevisiae* increases recombinant protein yield: the use of a respiratory strain as a microbial cell factory. Microb Cell Fact 9:47

44. van de Laar T, Visser C, Holster M, Lopez CG, Kreuning D, Sierkstra L, Lindner N, Verrips T (2007) Increased heterologous protein production by *Saccharomyces cerevisiae* growing on ethanol as sole carbon source. Biotechnol Bioeng 96:483–494

45. Gaspar ML, Aregullin MA, Jesch SA, Henry SA (2006) Inositol induces a profound alteration in the pattern and rate of synthesis and turnover of membrane lipids in *Saccharomyces cerevisiae*. J Biol Chem 281:22773–22785

46. Weuster-Botz D (2000) Experimental design for fermentation media development: statistical design or global random search? J Biosci Bioeng 90:473–483

47. Rezessy-Szabo JM, Nguyen QD, Hoschke A (2000) Optimisation of composition of media for the production of amylolytic enzymes by *Thermomyces lanuginosus* ATCC 34626. Food Technol Biotechnol 38:229–234

48. Ratnam BVV, Subba Rao S, Mendu DR, Narasimha Rao M, Ayyanna C (2005) Optimization of medium constituents and fermentation conditions for the production of ethanol from palmyra jaggery using response surface methodology. World J Microbiol Biotechnol 21:399–404

49. Olson BH, Johnson MJ (1949) Factors producing high yeast yields in synthetic media. J Bacteriol 57:235–246

50. Gasser B, Maurer M, Gach J, Kunert R, Mattanovich D (2006) Engineering of *Pichia pastoris* for improved production of antibody fragments. Biotechnol Bioeng 94:353–361

51. Travers KJ, Patil CK, Wodicka L, Lockhart DJ, Weissman JS, Walter P (2000) Functional and genomic analyses reveal an essential coordination between the unfolded protein response and ER-associated degradation. Cell 101:249–258

52. Perlman D, O'Brien E (1954) Characteristics of a cobalt tolerant culture of *Saccharomyces cerevisiae*. J Bacteriol 68:167–170

53. Bennett A, Rowe RI, Soch N, Eckhert CD (1999) Boron stimulates yeast (*Saccharomyces cerevisiae*) growth. J Nutr 129:2236–2238

Chapter 2

Which Yeast Species Shall I Choose? *Saccharomyces cerevisiae* Versus *Pichia pastoris* (Review)

Richard A.J. Darby, Stephanie P. Cartwright, Marvin V. Dilworth, and Roslyn M. Bill

Abstract

Having decided on yeast as a production host, the choice of species is often the first question any researcher new to the field will ask. With over 500 known species of yeast to date, this could pose a significant challenge. However, in reality, only very few species of yeast have been employed as host organisms for the production of recombinant proteins. The two most widely used, *Saccharomyces cerevisiae* and *Pichia pastoris*, are compared and contrasted here.

Key words: Yeast, Host cell, Recombinant protein production

1. Introduction

Yeast is a single-celled, eukaryotic microbe that can grow quickly in complex or defined media (doubling times are typically 2.5 h in glucose-containing media) and is easier and less expensive to use for recombinant protein production than insect or mammalian cells (1). These positive attributes make yeast suitable for use in formats ranging from multi-well plates, shake flasks and continuously stirred tank bioreactors to pilot plant and industrial scale reactors.

The most commonly employed species in the laboratory are *Saccharomyces cerevisiae* (also known as Baker's or Brewer's yeast) and some methylotrophic yeasts of the *Pichia* genus. In particular, *S. cerevisiae* and *P. pastoris* have both been genetically characterized (2–4) and shown to perform the posttranslational disulphide bond formation and glycosylation (5–7) that is crucial for the proper functioning of some recombinant proteins. However, it is impor-

Roslyn M. Bill (ed.), *Recombinant Protein Production in Yeast: Methods and Protocols*, Methods in Molecular Biology, vol. 866, DOI 10.1007/978-1-61779-770-5_2, © Springer Science+Business Media, LLC 2012

tant to note that yeast glycosylation does differ from that in mammalian cells: in *S. cerevisiae*, O-linked oligosaccharides contain only mannose moieties, whereas higher eukaryotic proteins have sialylated O-linked chains. Furthermore *S. cerevisiae* is known to hyperglycosylate N-linked sites, which can result in altered protein binding, activity, and potentially yield an altered immunogenic response in therapeutic applications (8). In *P. pastoris*, oligosaccharides are of much shorter chain length (9) and a strain has been reported that can produce complex, terminally sialylated or "humanized" glycoproteins (10).

Despite these potential limitations, recombinant pharmaceuticals including insulin (11), interferon-alpha-2a (Reiferon Retard®) and vaccines against hepatitis B virus (Hepavax-Gene and Engerix-B®) (12) and Human papilloma virus (Gardasil®) (13, 14) have all been produced in yeasts for commercial use, demonstrating the importance of yeast as a host organism to the pharmaceutical industry. The benefits and limitations of using *S. cerevisiae* and *P. pastoris* on a laboratory scale are addressed here and specific examples of their uses for the production of both soluble and membrane proteins are discussed.

2. Saccharomyces cerevisiae

S. cerevisiae is a single-celled, budding yeast, approximately 5–10 μm in size. Whilst it is commonly associated with the brewing and baking industries on account of its ability to produce ethanol and carbon dioxide, it is also the most widely studied eukaryotic organism. The USA's Food and Drug Administration (FDA) award of "generally recognized as safe" (GRAS) status to *S. cerevisiae* means that it is the most frequently used species of yeast for the production of many functional proteins. These include several soluble antibody fragments and fusions (15–19) as well as membrane protein drug targets such as G protein-coupled receptors (20–25), ABC transporters (9, 26) and drug resistance proteins (27).

2.1. Microbiology

The microbiology of *S. cerevisiae* is well understood and has been extensively reviewed elsewhere (28). In essence, it can grow both aerobically and anaerobically on a variety of carbon sources, is able to use ammonia or urea as a nitrogen source and also requires phosphorus and sulphur in its growth media. Certain metals such as calcium, iron, magnesium and zinc enhance its growth (29, 30): Tables 1 and 2 summarise typical growth media for *S. cerevisiae*. In culture, it has a relatively short generation time, doubling its cell density approximately every 1.5–2.5 h at its preferred growth temperature of 30°C.

The ability of *S. cerevisiae* to produce ethanol hints at its unusual metabolism: in most eukaryotes, oxygen depletion controls

Table 1
Composition of typical media for culturing *S. cerevisiae* and *P. pastoris*

S. cerevisiae medium	Components (L^{-1})
YPD (rich medium)	10 g bacto yeast extract 20 g bacto peptone 20 g glucose
YPG (rich medium with non-fermentable carbon source)	10 g bacto yeast extract 20 g bacto peptone 30 mL glycerol
CSM (complete synthetic medium)	1.7 g bacto yeast nitrogen base (without amino acids) 5 g ammonium sulphate 20 g glucose 100 mL 10× amino acid solution (see Table 2)
2× CBS (Centralbureau voor Schimmelcultures medium)	10 g ammonium sulphate 6 g potassium dihydrogen phosphate 1 g magnesium sulphate heptahydrate 20 g glucose 100 mL 1 M MES, pH 6 200 mL 10× amino acid solution (see Table 2) 2 mL vitamin solution (see Table 2) 2 mL trace element solution (see Table 2)

P. pastoris medium	Components (L^{-1})
BMGY (buffered glycerol-complex medium)	10 g bacto yeast extract 20 g bacto peptone 100 mL 10× YNB (see Table 2) 100 mL 1 M potassium phosphate (pH 6) 2 mL 500× biotin (see Table 2) 100 mL 10× glycerol (see Table 2)
BMMY (buffered methanol-complex medium)	10 g bacto yeast extract 20 g bacto peptone 100 mL 10× YNB (see Table 2) 100 mL 1 M potassium phosphate (see Table 2) 2 mL 500× biotin (see Table 2) 100 mL 10× methanol (see Table 2)
BSM (basal salts medium)	26.7 mL phosphoric acid 0.93 g calcium sulphate 18.2 g potassium sulphate 14.9 g magnesium sulphate heptahydrate 4.13 g potassium hydroxide 40 g glycerol 4.35 mL PTM$_1$ salts (see Table 2)

Table 2
Composition of stock solutions required to prepare media in Table 1

S. cerevisiae medium stocks	Components (L^{-1})
10× amino acid solution	200 mg L-adenine hemisulphate 200 mg L-arginine hydrochloride 200 mg L-histidine hydrochloride monohydrate 300 mg L-isoleucine 1000 mg L-leucine 300 mg L-lysine hydrochloride 200 mg L-methionine 500 mg L-phenylalanine 2000 mg L-threonine 200 mg L-tryptophan 300 mg L-tyrosine 200 mg L-uracil 1500 mg L-valine
Vitamin solution (filter sterilised)	0.05 g biotin 1 g calcium D-pantothenate 1 g nicotinic acid 25 g myo-inositol 1 g thiamine hydrochloride 1 g pyridoxol hydrochloride 0.2 g D-amino benzoic acid
Trace element solution	15 g EDTA 4.5 g zinc sulphate heptahydrate 1 g magnesium chloride tetrahydrate 0.3 g colbalt (II) chloride hexahydrate 0.3 g copper (II) sulphate pentahydrate 0.4 g sodium molybdate dihydrate 4.5 g calcium chloride dihydrate 3 g iron sulphate heptahydrate 1 g boric acid 0.1 g potassium iodide

P. pastoris medium stocks	Components (L^{-1})
10× YNB (filter sterilised)	34 g yeast nitrogen base without ammonium sulphate and amino acids 100 g ammonium sulphate
500× biotin (filter sterilised)	200 mg biotin
1 M potassium phosphate, pH 6	868 mL 1 M KH_2PO_4; 132 mL 1 M K_2HPO_4 (adjust to pH 6 with KOH and phosphoric acid)
10× glycerol (10%)	100 mL glycerol

(continued)

Table 2
(continued)

P. pastoris medium stocks	Components (L⁻¹)
10× methanol (5%; filter sterilised)	50 mL methanol
PTM₁ salts (filter sterilised)	6 g cupric sulphate pentahydrate 0.08 g sodium iodide 3 g manganese sulfate monohydrate 0.2 g sodium molybdate dihydrate 0.02 g boric acid 0.5 g cobalt chloride 20 g zinc chloride 65 g ferrous sulphate heptahydrate 0.2 g biotin 5 mL sulphuric acid

the switch from a respiratory to a fermentative metabolism, but in *S. cerevisiae* this switch also occurs in response to a change in the external concentration of a fermentable carbon source such as glucose (31). During the type of aerobic batch cultivation on glucose often performed on a laboratory scale, *S. cerevisiae* displays a biphasic growth pattern (Fig. 1). In the first respiro-fermentative phase, most of the glucose is converted to ethanol (32), which is subsequently consumed to produce carbon dioxide and water in the second phase. This has evolutionary advantages for *S. cerevisiae*, as the ethanol production phase is associated with a higher specific growth rate than the respiratory phase, providing a competitive advantage over other non-ethanol-producing organisms. Maximum recombinant protein yields are usually highest before yeast cells reach the end of this respiro-fermentative phase, before the so-called "diauxic shift" (Fig. 1) into the respiratory phase (33). Consequently, *S. cerevisiae* cells are typically harvested just before this diauxic shift in a protein production experiment, which can be readily assessed by monitoring the off-gas profile or the glucose concentration in the culture (24, 33). A respiratory strain of *S. cerevisiae*, TM6*, has been reported to have improved yield properties for both recombinant soluble and membrane proteins on account of its altered metabolism (24, 33). Its improved biomass yields, which are achieved at the expense of ethanol production, result in an increased volumetric yield of recombinant protein (34).

2.2. Genetics

S. cerevisiae was the first eukaryote to have its complete genome sequenced (2). The data are publicly available from the *Saccharomyces* Genome Database (SGD; http://www.yeastgenome.org), which is a scientific database of yeast molecular biology and genetics. The SGD provides detailed descriptions of the phenotypes of many mutant *S. cerevisiae* strains, many of which may have potential as

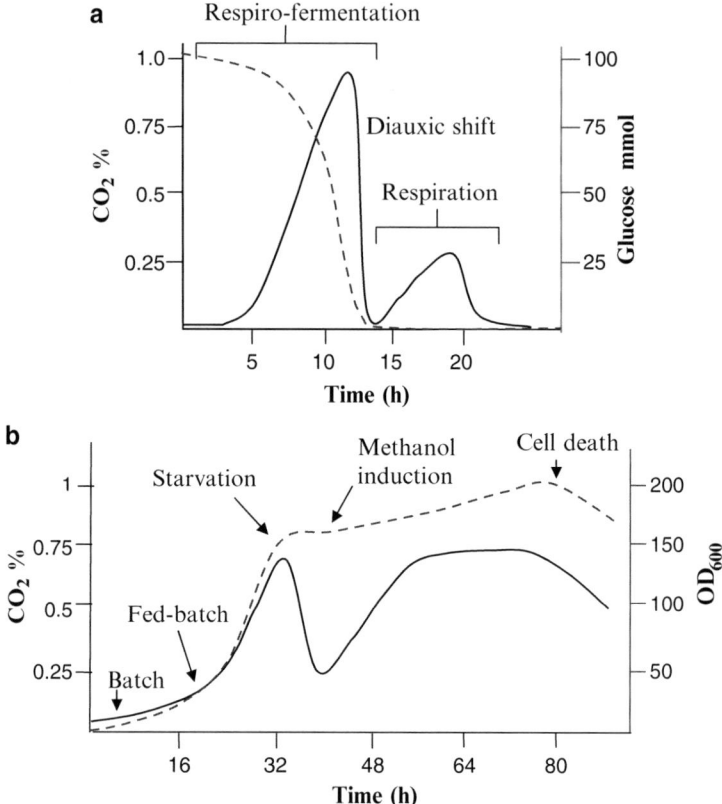

Fig. 1. (**a**) A representative CO_2 gas profile recorded in our laboratory, using a Tandem gas analyser, for a glucose-limited *S. cerevisiae* culture grown in a 2 L Applikon bioreactor. The *solid line* shows the production of CO_2 (%) and the *dashed line* shows the depletion of glucose (mmol). The respiro-fermentation and respiration phases are indicated, as is the diauxic shift between the two phases. (**b**) A CO_2 gas profile recording for a *P. pastoris* glycerol fed-batch bioreactor culture induced at 28 h with a limiting amount of methanol. The *solid line* shows the production of CO_2 (%) and the *dashed line* shows the optical density of the culture.

protein production hosts, with links to supporting literature. Importantly, the SGD is fully annotated and is continually updated.

Resources such as the SGD have facilitated an increased understanding of *S. cerevisiae* in general and, more specifically in the context of this discussion, of recombinant protein production in yeast cells. For example, since the production of high yields of functional recombinant proteins, particularly membrane proteins, remains a significant challenge, we and others (35, 36) have focussed on characterising the cellular response to recombinant protein production. Identifying specific genes that exhibit an altered transcriptional profile, when the cell produces high yields of functional recombinant proteins, has been used to guide subsequent engineering of high-yielding strains (24, 33, 37). This approach is

Table 3
Examples of expression vectors available for *S. cerevisiae* and *P. pastoris*

Vector	Yeast	Selection marker	Promoter	Expression	Episomal or integrative
pYX212	*S. cerevisiae*	*URA3*	*TPI1*	Constitutive	Episomal
pYX222	*S. cerevisiae*	*HIS3*	*TPI1*	Constitutive	Episomal
pYES2	*S. cerevisiae*	*URA3*	*GAL1*	Inducible	Episomal
pVTU260	*S. cerevisiae*	*URA3*	*ADH1*	Constitutive	Episomal
YEpCTHS	*S. cerevisiae*	*Ampicillin*	*CUP1*	Inducible	Episomal
pPICZ	*P. pastoris*	*Zeocin*	*AOX1*	Inducible	Integrative
pGAPZ	*P. pastoris*	*Zeocin*	*GAP*	Constitutive	Integrative
pBGP1	*P. pastoris*	*Zeocin*	*GAP*	Constitutive	Episomal
pGAPZ-E	*P. pastoris*	*Zeocin*	*GAP*	Constitutive	Episomal

strengthened by the availability of a complete set of single, non-essential gene deletion strains (EUROSCARF) as well as a strain collection of tetracycline-regulated essential genes (Open Biosystems). We demonstrated that increased yields of recombinant proteins can be achieved when specific members of these collections are used as host organisms (37). This permits the production of a recombinant protein to be compared in multiple strains simultaneously whilst gaining an improved knowledge of the molecular pathways involved in producing the protein.

2.3. Molecular Biology

The DNA sequence encoding the target protein of interest is typically amplified via PCR either from genomic DNA or cDNA and then cloned into a suitable expression plasmid with or without signal sequences and fusion partners (see Chapters 3 and 4). Plasmids for *S. cerevisiae* can be sub-divided into three categories: low-copy number replicating plasmids, multi-copy number replicating plasmids and integrative plasmids (38, 39). These can, in turn, contain a range of different promoters of varying strengths that are either inducible or constitutive (Table 3). This means that a variety of different options can be tested in order to optimize the most suitable regime for recombinant protein production. However, it is important to ensure the stability of any transformants generated. For example, autonomous plasmids can be relatively unstable, yielding a heterogenous population of transformants that routinely require screening for the desired expression level, as well as being prone to genetic loss upon cell division. Furthermore, high-copy number plasmids may also result in expression levels that can overwhelm the host's post-translational and secretory pathways, yielding misfolded and degraded protein. It may be possible to overcome some of these problems by integrating the expression cassette into the genome, thereby increasing its genetic stability (38, 39).

3. Pichia pastoris

P. pastoris is used increasingly as the host cell of choice because of its ability to produce high yields of properly folded proteins in exceptionally high-density cultures. To achieve the highest possible biomass yields, it must be cultured in fully controlled, continuously stirred tank bioreactors (40). Its emergence as an alternative to *S. cerevisiae* is exemplified by the variety of heterologous proteins it has been used to produce in high yields, ranging from tetanus toxin and mouse epidermal growth factor (41–43) to membrane proteins including human ABC transporters, aquaporins and tetraspanins (44–46). A number of different protease-deficient strains are also available (SMD1163, 1165 and 1168) that have been shown to exhibit reduced proteolysis of some recombinant proteins (47–50).

3.1. Microbiology

P. pastoris has a respiratory metabolism (Fig. 1) and can be cultured to exceptionally high cell densities (hundreds of grams per litre) on glycerol-containing media (Tables 1 and 2), often yielding a culture resembling a paste at the end of an experiment (43, 51). The development of a respiratory strain of *S. cerevisiae* (TM6*) (24, 52) has gone someway to permitting similarly high cell density cultures of *S. cerevisiae*. However, while high cell density cultures are very attractive for increasing volumetric yields, productivity does not necessarily increase linearly with increased biomass yields, and in some situations may actually decrease (40, 53). For this reason, the ability to increase "per cell" yields is an area of active research, not only in *P. pastoris* but also in other host cells (34).

P. pastoris is a methylotroph, which has two endogenous copies of the *AOX* gene; *AOX1* expression accounts for more than 90% of the enzyme in the cell whilst *AOX2* expression constitutes less than 10% (see Chapter 15). It is these genes that permit the utilisation of methanol as a carbon source: using the *Pichia* Expression Kit marketed by Invitrogen Corporation, it is possible to use the *AOX* promoter to control heterologous gene expression, which is induced in the presence of methanol and repressed by glucose or glycerol (51). The careful control of the methanol induction regime is central to increasing yields per cell (40). Methanol metabolism is highly dependent on oxygen availability within the culture and it is widely accepted that the dissolved oxygen concentration (DO) should be maintained above 20% (54). This regime has been successfully employed to produce many different soluble and membrane proteins (40, 44, 46, 53, 55–57).

Despite methanol induction being robust and tightly regulated, the potential risks associated with methanol use, such as its toxicity to cells at concentrations above 5 g/L and its volatility, have led researchers to investigate alternative promoters that do not require the use of methanol (58). Constitutive expression of heterologous genes can be achieved when cloned downstream of the glyceraldehyde 3-phosphate dehydrogenase (*GAP*) promoter (59, 60), whilst strong induction via the formaldehyde dehydrogenase (*FLD1*) promoter has also been reported in the presence of methylamine as well as methanol (61).

Like *S. cerevisiae*, *P. pastoris* is capable of producing disulphide-bonded and glycosylated proteins (62). However, the glycosylation pattern is different in *P. pastoris* compared with *S. cerevisiae* (63): in *P. pastoris* N-linked oligosaccharides are usually no more than 20 residues in length compared with 50–150 residues in *S. cerevisiae*. In addition, *P. pastoris* lacks the mannosyl transferase which yields immunogenic α-1, 3-linked mannosyl terminal linkages in *S. cerevisiae* (64).

3.2. Genetics

P. pastoris is not considered to be as genetically amenable as *S. cerevisiae*, despite the fact that a genomic sequence of the GS115 strain has been commercially available since the mid-2000s. This is partly on account of a restrictive user contract that required all sequence information to be confidentially maintained. Despite this, genetic advances have been made as highlighted by the development of a "humanized" *P. pastoris* strain (10) capable of replicating the most essential glycosylation pathways found in mammalian cells and permitting the production of active recombinant erythropoietin. The open-access publication of the GS115 (3) and DSMZ 70382 (4) genomes in 2009 and their respective annotation at http://bioinformatics.psb.ugent.be/webtools/bogas/ and http://www.pichiagenome.org should make a significant impact in this area, as both sites permit free access to view the genomic sequences and use sequence resource software (65).

3.3. Molecular Biology

As for *S. cerevisiae*, the target protein's DNA sequence is often PCR amplified from genomic DNA or cDNA and cloned into an expression plasmid with or without signal sequences and fusion partners. The most widely used *P. pastoris* expression vectors are designed to be maintained as stable, integrative elements in its genome (see Chapter 3). Examples include Invitrogen's pPIC and pGAPZ series of vectors (Table 3). However, tranformants often exhibit heterogeneous expression levels, and this necessitates the screening of many colonies to isolate high-yielding clones (see Chapter 7). The limited number of episomal plasmids for *P. pastoris* to date has been predominantly due to plasmid instability during replication (66). Those that are available (Table 3) often utilize the constitutive *GAP* promoter (67, 68) and require the addition of selective antibiotic to maintain the vector. With the advent of open access genomic data it is hoped that there will be an increase in the number of episomal vectors that contain different auxotrophic markers for selection.

4. Which Species Should I Choose?

There are benefits and drawbacks to using both *S. cerevisiae* and *P. pastoris* as hosts for recombinant protein production. For the production of secreted proteins, *P. pastoris* may be the best choice on account of its limited endogenous protein secretion and the number of different protease-deficient strains available. However, the full benefit of *P. pastoris* is achievable only if it is cultured under strictly defined conditions, usually only possible in continuously stirred tank bioreactors. Therefore the optimal use of *P. pastoris* may require a more long-term investment of time and equipment resources than for *S. cerevisiae*. In contrast *S. cerevisiae* provides a much wider range of resources (both strains and expression vectors) and is supported by a much more extensive literature than *P. pastoris*. Consequently, projects requiring a range of strains may benefit from using *S. cerevisiae* as the host. In our laboratory, we often start with *P. pastoris* and if the production is not straightforward, turn to *S. cerevisiae* to troubleshoot, thereby benefitting from the best attributes of the two hosts.

References

1. Bill RM (2001) Yeast – a panacea for the structure-function analysis of membrane proteins? Curr Genet 40:157–171
2. Goffeau A, Barrell BG, Bussey H, Davis RW, Dujon B, Feldmann H, Galibert F, Hoheisel JD, Jacq C, Johnston M, Louis EJ, Mewes HW, Murakami Y, Philippsen P, Tettelin H, Oliver SG (1996) Life with 6000 genes. Science 274:546–567
3. De Schutter K, Lin YC, Tiels P, Van Hecke A, Glinka S, Weber-Lehmann J, Rouze P, de Peer YV, Callewaert N (2009) Genome sequence of the recombinant protein production host *Pichia pastoris*. Nat Biotechnol 27:561–566
4. Mattanovich D, Graf A, Stadlmann J, Dragosits M, Redl A, Maurer M, Kleinheinz M, Sauer M, Altmann F, Gasser B (2009) Genome, secretome and glucose transport highlight unique features of the protein production host *Pichia pastoris*. Microb Cell Fact 8:29
5. Demain AL, Vaishnav P (2009) Production of recombinant proteins by microbes and higher organisms. Biotechnol Adv 27:297–306
6. Ferrer-Miralles N, Domingo-Espin J, Corchero JL, Vazquez E, Villaverde A (2009) Microbial factories for recombinant pharmaceuticals. Microb Cell Fact 8:17
7. Jigami Y (2008) Yeast glycobiology and its application. Biosci Biotechnol Biochem 72:637–648
8. Gerngross TU (2004) Advances in the production of human therapeutic proteins in yeasts and filamentous fungi. Nat Biotechnol 22:1409–1414
9. Bretthauer RK, Castellino FJ (1999) Glycosylation of *Pichia pastoris* – derived proteins. Biotechnol Appl Biochem 30:193–200
10. Hamilton SR, Davidson RC, Sethuraman N, Nett JH, Jiang YW, Rios S, Bobrowicz P,

Stadheim TA, Li HJ, Choi BK, Hopkins D, Wischnewski H, Roser J, Mitchell T, Strawbridge RR, Hoopes J, Wildt S, Gerngross TU (2006) Humanization of yeast to produce complex terminally sialylated glycoproteins. Science 313:1441–1443

11. Kjeldsen T, Ludvigsen S, Diers I, Balschmidt P, Sørensen AR, Kaarsholm NC (2002) Engineering-enhanced protein secretory expression in yeast with application to insulin. J Biol Chem 277:18245–18248

12. Andre FE, Safary A (1987) Summary of clinical findings on engerix-b, a genetically engineered yeast-derived Hepatitis-B vaccine. Postgrad Med J 63:169–178

13. Siddiqui MAA, Perry CM (2006) Human papillomavirus quadrivalent (types 6, 11, 16, 18) recombinant vaccine (Gardasil (R)): profile report. BioDrugs 20:313–316

14. Siddiqui MAA, Perry CM (2006) Human papillomavirus quadrivalent (types 6, 11, 16, 18) recombinant vaccine (Gardasil (R)). Drugs 66:1263–1271

15. Gasser B, Mattanovich D (2007) Antibody production with yeasts and filamentous fungi: on the road to large scale? Biotechnol Lett 29:201–212

16. Hackel BJ, Huang DG, Buboz JC, Wang XX, Shusta EV (2006) Production of soluble and active transferrin receptor-targeting single-chain antibody using *Saccharomyces cerevisiae*. Pharm Res 23:790–797

17. Evans L, Hughes M, Waters J, Cameron J, Dodsworth N, Tooth D, Greenfield A, Sleep D (2010) The production, characterisation and enhanced pharmacokinetics of scFv-albumin fusions expressed in *Saccharomyces cerevisiae*. Protein Expr Purif 73:113–124

18. Frenken LGJ, van der Linden RHJ, Hermans PWJJ, Bos JW, Ruuls RC, de Geus B, Verrips CT (2000) Isolation of antigen specific Llama V-HH antibody fragments and their high level secretion by *Saccharomyces cerevisiae*. J Biotechnol 78:11–21

19. Edqvist J, Keranen S, Penttila M, Straby KB, Knowles JKC (1991) Production of functional Igm Fab fragments by *Saccharomyces cerevisiae*. J Biotechnol 20:291–300

20. Liitti S, Matikainen MT, Scheinin M, Glumoff T, Goldman A (2001) Immunoaffinity purification and reconstitution of human alpha(2)-adrenergic receptor subtype C2 into phospholipid vesicles. Protein Expr Purif 22:1–10

21. Huang HJ, Liao CF, Yang BC, Kuo TT (1992) Functional expression of rat M5 muscarinic acetylcholine-receptor in yeast. Biochem Biophys Res Commun 182:1180–1186

22. Price LA, Strnad J, Pausch MH, Hadcock JR (1996) Pharmacological characterization of the rat A(2a) adenosine receptor functionally coupled to the yeast pheromone response pathway. Mol Pharmacol 50:829–837

23. Joubert O, Nehme R, Bidet M, Mus-Veteau I (2010) Heterologous expression of human membrane receptors in the yeast *Saccharomyces cerevisiae*, heterologous expression of membrane proteins. Methods Mol Biol 601:87–103

24. Ferndahl C, Bonander N, Logez C, Wagner R, Gustafsson L, Larsson C, Hedfalk K, Darby RAJ, Bill RM (2010) Increasing cell biomass in *Saccharomyces cerevisiae* increases recombinant protein yield: the use of a respiratory strain as a microbial cell factory. Microb Cell Fact 9:47

25. Kapat A, Jaakola VP, Heimo H, Liitti S, Heikinheimo P, Glumoff T, Goldman A (2000) Production and purification of recombinant human alpha 2C2 adrenergic receptor using *Saccharomyces cerevisiae*. Bioseparation 9:167–172

26. Duman JG, Miele RG, Liang H, Grella DK, Sim KL, Castellino FJ, Bretthauer RK (1998) O-Mannosylation of *Pichia pastoris* cellular and recombinant proteins. Biotechnol Appl Biochem 28:39–45

27. Miele RG, Castellino FJ, Bretthauer RK (1997) Characterization of the acidic oligosaccharides assembled on the *Pichia pastoris*-expressed recombinant kringle 2 domain of human tissue-type plasminogen activator. Biotechnol Appl Biochem 26:79–83

28. Barnett JA, Barnett L (2011) Yeast research: a historical overview. ASM, Herndon, VA

29. Treco DA, Lundblad V (2001) Preparation of yeast media. Curr Protoc Mol Biol Chapter 13, Unit13.1

30. Curran BP, Bugeja V (2006) Basic investigations in *Saccharomyces cerevisiae*. Methods Mol Biol 313:1–13

31. Bonander N, Ferndahl C, Mostad P, Wilks MD, Chang C, Showe L, Gustafsson L, Larsson C, Bill RM (2008) Transcriptome analysis of a respiratory *Saccharomyces cerevisiae* strain suggests the expression of its phenotype is glucose insensitive and predominantly controlled by *Hap4*, *Cat8* and *Mig1*. BMC Genomics 9:365

32. Verduyn C, Zomerdijk TPL, Dijken JP, Scheffers WA (1984) Continuous measurement of ethanol production by aerobic yeast suspensions with an enzyme electrode. Appl Microbiol Biotechnol 19:181–185

33. Bonander N, Hedfalk K, Larsson C, Mostad P, Chang C, Gustafsson L, Bill RM (2005) Design of improved membrane protein production

experiments: quantitation of the host response. Protein Sci 14:1729–1740

34. Bawa Z, Bland CE, Bonander N, Bora N, Cartwright SP, Clare M, Conner MT, Darby RA, Dilworth MV, Holmes WJ, Jamshad M, Routledge SJ, Gross SR, Bill RM (2011) Understanding the yeast host cell response to recombinant membrane protein production. Biochem Soc Trans 39:719–723

35. Wang H, Prorok M, Bretthauer RK, Castellino FJ (1997) Serine-578 is a major phosphorylation locus in human plasma plasminogen. Biochemistry 36:8100–8106

36. Ren J, Castellino FJ, Bretthauer RK (1997) Purification and properties of alpha-mannosidase II from Golgi-like membranes of baculovirus-infected *Spodoptera frugiperda* (IPLB-SF-21AE) cells. Biochem J 324:951–956

37. Bonander N, Darby RAJ, Grgic L, Bora N, Wen J, Brogna S, Poyner DR, O'Neill MAA, Bill RM (2009) Altering the ribosomal subunit ratio in yeast maximizes recombinant protein yield. Microb Cell Fact 8:10

38. Schneider JC, Guarente L (1991) Vectors for expression of cloned genes in yeast: regulation, overproduction and underproduction. Methods Enzymol 194:373–388

39. Zhang Z, Moo-Young M, Chisti Y (1996) Plasmid stability in recombinant *Saccharomyces cerevisiae*. Biotechnol Adv 14:401–435

40. Holmes WJ, Darby RAJ, Wilks MDB, Smith R, Bill RM (2009) Developing a scalable model of recombinant protein yield from *Pichia pastoris*: the influence of culture conditions, biomass and induction regime. Microb Cell Fact 8:35

41. Clare JJ, Rayment FB, Ballantine SP, Sreekrishna K, Romanos MA (1991) High-level expression of tetanus toxin fragment C in *Pichia pastoris* strains containing multiple tandem integrations of the gene. Biotechnology 9:455–460

42. Clare JJ, Romanos MA, Rayment FB, Rowedder JE, Smith MA, Payne MM, Sreekrishna K, Henwood CA (1991) Production of mouse epidermal growth-factor in yeast – high-level secretion using *Pichia pastoris* strains containing multiple gene copies. Gene 105:205–212

43. Macauley-Patrick S, Fazenda ML, McNeil B, Harvey LM (2005) Heterologous protein production using the *Pichia pastoris* expression system. Yeast 22:249–270

44. Rosenberg MF, Bikadi Z, Chan J, Liu XP, Ni ZL, Cai XK, Ford RC, Mao QC (2010) The human breast cancer resistance protein (BCRP/ABCG2) shows conformational changes with mitoxantrone. Structure 18:482–493

45. Urbatsch IL, Wilke-Mounts S, Gimi K, Senior AE (2001) Purification and characterization of

n-glycosylation mutant mouse and human p-glycoproteins expressed in *Pichia pastoris* cells. Arch Biochem Biophys 388:171–177

46. Jamshad M, Rajesh S, Stamataki Z, McKeating JA, Dafforn T, Overduin M, Bill RM (2008) Structural characterization of recombinant human CD81 produced in *Pichia pastoris*. Protein Expr Purif 57:206–216

47. Grunewald S, Haase W, Molsberger E, Michel H, Reilander H (2004) Production of the human D-2S receptor in the methylotrophic yeast *P. pastoris*. Receptors Channels 10:37–50

48. Shi XL, Feng MQ, Shi J, Shi ZHA, Zhong JA, Zhou P (2007) High-level expression and purification of recombinant human catalase in *Pichia pastoris*. Protein Expr Purif 54:24–29

49. Ogunjimi AA, Chandler JM, Gooding CM, Recinos A, Choudary PV (1999) High-level secretory expression of immunologically active intact antibody from the yeast *Pichia pastoris*. Biotechnol Lett 21:561–567

50. Andre N, Cherouati N, Prual C, Steffan T, Zeder-Lutz G, Magnin T, Pattus F, Michel H, Wagner R, Reinhart C (2006) Enhancing functional production of G protein-coupled receptors in *Pichia pastoris* to levels required for structural studies via a single expression screen. Protein Sci 15:1115–1126

51. Cregg JM, Cereghino JL, Shi JY, Higgins DR (2000) Recombinant protein expression in *Pichia pastoris*. Mol Biotechnol 16:23–52

52. Otterstedt K, Larsson C, Bill RM, Stahlberg A, Boles E, Hohmann S, Gustafsson L (2004) Switching the mode of metabolism in the yeast *Saccharomyces cerevisiae*. EMBO Rep 5:532–537

53. Singh S, Hedley D, Kara E, Gras A, Iwata S, Ruprecht J, Strange PG, Byrne B (2010) A purified C-terminally truncated human adenosine A2A receptor construct is functionally stable and degradation resistant. Protein Expr Purif 74:80–87

54. Li PZ, Anumanthan A, Gao XG, Ilangovan K, Suzara VV, Duzgunes N, Renugopalakrishnan V (2007) Expression of recombinant proteins in *Pichia pastoris*. Appl Biochem Biotechnol 142:105–124

55. Yinliang C, Cino J, Hart G, Freedman D, White C, Komives EA (1997) High protein expression in fermentation of recombinant *Pichia pastoris* by a fed-batch process. Process Biochem 32:107–111

56. Jin H, Liu G, Ye X, Duan Z, Li Z, Shi Z (2010) Enhanced porcine interferon-α production by recombinant *Pichia pastoris* with a combinational control strategy of low induction temperature and high dissolved oxygen concentration. Biochem Eng J 52:91–98

57. Fraser NJ (2006) Expression and functional purification of a glycosylation deficient version of the human adenosine 2a receptor for structural studies. Protein Expr Purif 49:129–137

58. Cos O, Ramon R, Montesinos JL, Valero F (2006) Operational strategies, monitoring and control of heterologous protein production in the methylotrophic yeast *Pichia pastoris* under different promoters: a review. Microb Cell Fact 5:17

59. Waterham HR, Digan ME, Koutz PJ, Lair SV, Cregg JM (1997) Isolation of the *Pichia pastoris* glyceraldehyde-3-phosphate dehydrogenase gene and regulation and use of its promoter. Gene 186:37–44

60. Kim SJ, Lee JA, Kim YH, Song BK (2009) optimization of the functional expression of coprinus cinereus peroxidase in *Pichia pastoris* by varying the host and promoter. J Microbiol Biotechnol 19:966–971

61. Resina D, Cos O, Ferrer P, Valero F (2005) Developing high cell density fed-batch cultivation strategies for heterologous protein production in *Pichia pastoris* using the nitrogen source-regulated FLD1 promoter. Biotechnol Bioeng 91:760–767

62. Daly R, Hearn MTW (2005) Expression of heterologous proteins in *Pichia pastoris*: a useful experimental tool in protein engineering and production. J Mol Recognit 18:119–138

63. Dale C, Allen A, Fogerty S (1999) *Pichia pastoris*: a eukaryotic system for the large-scale production of biopharmaceuticals. Biopharm 12:36–40

64. Cregg JM, Vedvick TS, Raschke WC (1993) Recent advances in the expression of foreign genes in *Pichia pastoris*. Nat Biotechnol 11:905–910

65. Mattanovich D, Callewaert N, Rouze P, Lin YC, Graf A, Redl A, Tiels P, Gasser B, De Schutter K (2009) Open access to sequence: browsing the *Pichia pastoris* genome. Microb Cell Fact 8:53

66. Cregg JM, Barringer KJ, Hessler AY, Madden KR (1985) *Pichia-pastoris* as a host system for transformations. Mol Cell Biol 5:3376–3385

67. Lee CC, Williams TG, Wong DWS, Robertson GH (2005) An episomal expression vector for screening mutant gene libraries in *Pichia pastoris*. Plasmid 54:80–85

68. Choi SG, Hong IP, Anderson S (2006) Evaluation of a new episomal vector based on the GAP promoter for structural genomics in *Pichia pastoris*. J Microbiol Biotechnol 16:1362–1368

Chapter 3

Preparation of *Pichia pastoris* Expression Plasmids

Christel Logez, Fatima Alkhalfioui, Bernadette Byrne, and Renaud Wagner

Abstract

When planning any heterologous expression experiment, the very first critical step is related to the design of the overall strategy, hence to the selection of the most adapted expression vector. The very flexible *Pichia pastoris* system offers a broad range of possibilities for the production of secreted, endogenous or membrane proteins thanks to a combination of various plasmid backbones, selection markers, promoters and fusion sequences introduced into dedicated host strains. The present chapter provides some guidelines on the choice of expression vectors and expression strategies. It also brings the reader a complete toolbox from which plasmids and fusion sequences can be picked and assembled to set up appropriate expression vectors. Finally, it provides standard starting protocols for the preparation of the selected plasmids and their use for host strain transformation.

Key words: Plasmid, Expression, Purification/detection tag, Promoter, *Pichia pastoris* cell strains

1. Introduction

Hundreds of proteins of various types, origins and functions have been produced in *Pichia pastoris* for many purposes and applications. Conveniently, a large set of representative examples have been listed in authoritative reviews and can advantageously serve the reader as points of reference (1–4). In these numerous studies, high yields are very often dependent on several parameters including the choice of the expression vector, the optimal gene sequence, the nature and site of insertion of any fusion tags, and the transformation and selection strategies. Thus, while no standards exist to predict which combination will enable the successful production of a given protein, we instead propose the following steps that may help in determining the appropriate tools and methods to start with, and where to find them.

Roslyn M. Bill (ed.), *Recombinant Protein Production in Yeast: Methods and Protocols*, Methods in Molecular Biology, vol. 866, DOI 10.1007/978-1-61779-770-5_3, © Springer Science+Business Media, LLC 2012

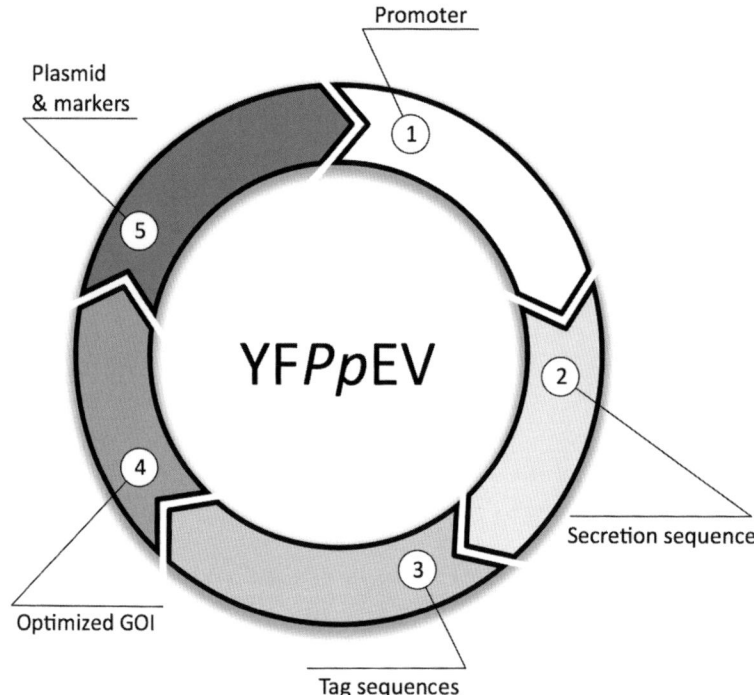

Fig. 1. Your Favourite *P. pastoris* Expression Vector (YF*Pp*EV). The five different boxes represent the main elements that must be considered when building an expression vector. *GOI* gene of interest.

1.1. What Plasmid to Select?

Except for a limited number of autoreplicative plasmids that are not yet frequently employed (5–8), the usual expression vectors are designed to be maintained as integrative elements in the genome of *P. pastoris* (see Subheading 1.6. below). They are built on a classical *E. coli*/yeast shuttle model with components required for *E. coli* amplification (classically one origin of replication and one antibiotic selection marker) and specific elements for heterologous gene expression in *P. pastoris*. These typically include selectable auxotrophy markers and/or antibiotic-resistance bacterial genes, a range of promoter and terminator sequences, a cloning cassette and supplementary fusion sequences that can be added for improving the secretion, detection and purification of the recombinant proteins. Each is schematically represented as a building block in Fig. 1, and these are further discussed in the following sections.

1.2. What Promoter to Use?

P. pastoris harbours several strong or weaker promoters that can be exploited to drive heterologous expression of recombinant genes, both in an inducible or constitutive fashion (see Table 1). Inducible expression is usually the preferred strategy since it allows a convenient control of the experimental conditions applied before expression and is ideally adapted for the production of proteins that are toxic to the host. *P. pastoris* offers a panel of promoters that can be induced in the presence of various carbon or nitrogen sources (9),

Table 1
P. pastoris promoters for recombinant expression

Name	Original gene	Nature (expression condition)	Strength	References
P_{AOX1}	Alcohol oxidase I	Inducible (*methanol*)	Strong	(29)
P_{AOX2}	Alcohol oxidase II	Inducible (*methanol*)	Strong	(30)
P_{GAP}	Glyceraldehyde-3-phosphate dehydrogenase	Constitutive (*glucose, glycerol*)	Strong	(31)
P_{FLD1}	Formaldehyde dehydrogenase	Inducible (*methanol, methylamine*)	Strong	(32)
P_{DHAS}	Dihydroxyacetone synthase	Inducible (*methanol*)	Weak	(33)
P_{ICL}	Isocitrate lyase	Inducible (*ethanol*)	Not determined	(34)
P_{YPT1}	GTP-binding protein	Constitutive (*glucose, methanol, mannitol*)	Weak	(35)
P_{TEF1}	Translation elongation factor 1α	Constitutive (*glucose*)	Strong	(36)
P_{PHO89}	Sodium phosphate symporter	Inducible (*phosphate starvation*)	Strong	(37)
P_{PEX8}	Peroxisomal matrix protein	Inducible (*methanol*)	Weak	(38)

the promoter P_{AOX1} from the alcohol oxidase encoding gene (*AOX1*) being predominantly employed. This promoter is tightly repressed by glucose and strongly induced by methanol (10) allowing the cells to use methanol as the sole carbon source. A P_{AOX1} synthetic promoter library was recently developed revealing enhanced P_{AOX1} variants that resulted in high levels of a recombinant GFP (11). There are numerous cases, however, where constitutive expression performs as well as inducible expression, in particular when using the strong glyceraldehyde-3-phosphate dehydrogenase P_{GAP} promoter (9, 12). In addition, constitutive expression is more straightforward to manage since no switch of carbon source is required, which is particularly convenient when running fermentation procedures.

1.3. Do I Need a Secretion Signal?

This is a non-trivial question since the choice of intracellular or extracellular localisation can have a direct impact both on yield and integrity of the recombinant protein, as well as on the procedures required for isolation. Secreting the recombinant proteins outside the cell has several advantages: soluble protein production may be

Table 2
Signal sequences used for extracellular expression

Name	Protein signal sequence	Sequence structure	References
α factor	*S. cerevisiae* α mating factor	85 residues 3 N-glycosylation sites Kex2/Ste13 processing sites	(39)
PHO1	*P. pastoris* acid phosphatase	15 residues 6 N-glycosylation sites	(40)
SUC2	*P. pastoris* invertase	19 residues	(39, 41)
PHA-E	*Phaseolus vulgaris* phytohaemagglutinin	21 residues	(42)
KILM1	Killer toxin type I	44 residues	(42)
pGKL	128-kDa killer protein	29 residues Kex2 processing site	(43, 44)
CLY	Chicken lysozyme signal peptide	18 residues	(45)
CLY-L8	Engineered leucine-rich signal peptide	16 residues	(45)

induced for longer periods of time since they are not accumulating in the limited volume of the cytoplasm where they might become toxic for the host. This can lead to an increased recombinant yield. Furthermore, no cell-lysis step is required and secreted proteins can be recovered directly from the culture medium, which contains far fewer contaminating proteins than the cells therefore simplifying the purification process. One limitation is the frequent degradation of the secreted proteins by extracellular proteases and proteases released from lysed cells. In addition, proteins that are not naturally secreted may not be properly folded outside the cell. In this regard, intracellular production is a valuable alternative (13, 14).

When opting for a secretion strategy, the target protein needs to be identified as secreted by the presence of a signal sequence (see Table 2). Successful secretion of many proteins from *P. pastoris* has been reported using a range of different signal sequences. These include a protein's native secretion signal, the *Saccharomyces cerevisiae* α-mating factor prepro leader sequence (α-MF), the *P. pastoris* acid phosphatase (*PHO1*) signal sequence and the invertase (*SUC2*) signal sequence (see ref. 1 for an extensive list). Further information on the range and use of prepro peptides can be found in ref. (15) and this may provide a useful resource for selecting a signal sequence.

In the case of integral membrane proteins, adding a secretion sequence may be highly beneficial for expression in *P. pastoris*. Such an approach has proved highly effective for the production of GPCRs (16, 17). However, the presence of the signal sequence has variable effects on the expression of other membrane protein-encoding genes. In the case of aquaporins for instance, where the amino- and carboxy-termini are both located intracellularly, protein production has been evaluated with or without a fused signal sequence. In both cases high yields of high-quality protein suitable for structural studies were obtained (18, 19).

1.4. What Kind of Additional Sequences Do I Need?

Whatever the objectives to be achieved when producing a protein with *P. pastoris* (biochemical and/or biophysical characterisation, structural studies, pharmaceutical or food production), recovering the protein in its most native form is generally mandatory. There are many examples of expression and subsequent isolation of untagged proteins requiring development of specific and often tedious purification procedures. Alternatively, adding epitope tags allows detection, and isolation of the target protein using generic techniques and tools. An ideal tag should not only (1) exert a minimal effect on the tertiary structure and the biological activity of the protein it is fused to but should also (2) allow a one-step adsorption purification, (3) be easily and specifically removed to produce the native protein and (4) be applicable to a number of different proteins. While it is difficult to decide on the best fusion sequence and position to be used for a specific protein, we present in Table 3 a list of tags and protease cleavage sites to release them that have proven helpful for the production of proteins in *P. pastoris* (20).

1.5. How Can I Optimise the Sequence of My Protein-Encoding Gene?

Even if recombinant genes are most often cloned and expressed in their native form, several adjustments in their sequence can be made to best fit the transcription and translation machineries of the yeast and very often result in dramatic improvement of the protein yields. The sequence parameters that were notably shown to positively influence the expression levels include (1) an optimal translation initiation sequence (the yeast consensus is A/YAA/UAAUGUCU), (2) an adaptation to the codon usage of yeasts (21, 22), (3) an increase of the GC-content (22, 23), (4) a decreased occurrence of AT-rich regions (24) and (5) an adapted isoelectric point of the protein (24). With the very recent release of the whole genome sequence of *P. pastoris* (25), even more accurate sequence optimisations are now possible.

Answering this series of questions should help the researcher to assemble the most suitable vector(s) for production of their target protein in *P. pastoris*. A significant set of plasmids is commercially available from Invitrogen (see the Protein and Expression section in http://www.invitrogen.com and Table 4) that could either be used as is or further engineered to best suit the selected strategy.

1.6. My Construct Is Now Ready, How Do I Proceed with P. pastoris Transformation?

As for many other yeasts, transformation of *P. pastoris* is straightforward. Several robust methods are available, based on either chemically competent (spheroplasts, PEG1000, LiCl) or electrocompetent cells. Moreover, these protocols are well described and can be easily found on numerous Web sites: convenient *Pichia* manuals can be downloaded from http://www.invitrogen.com.

Table 3
Additional tag sequences and protease cleavage sites

(a) Additional tag sequences

Name	Peptidic sequence	Size
His-tag	HHHHHH(HHHH)	6–10 aa.
Flag-tag	DYKDDDDK	8 aa.
Strep II tag	WSHPQFEK	8 aa.
1D4 antibody recognition sequence	TETSQVAPA	9 aa.
HA epitope tag	YPYDVPDYAS	10 aa.
c-Myc epitope	EQKLISEEDL	11 aa.
V5 epitope	GKPIPNPLLGLDST	14 aa.
Calmodulin-binding peptide	KRRWKKNFIAVSAANRFKKISSSGAL	26 aa.
Bio-tag	Peptide	85 aa.
GFP	Protein	35 kDa

(b) Protease cleavage sites

Name	Recognised sequence	Protease size (kDa)
Enterokinase	DDDDK	26
Tobacco etch virus (TEV) protease	ENLYFQS	28
Thrombin	LVPRGS/F	35
Factor Xa.	IE/DGR	43

The number of strains usually employed for heterologous expression is rather limited (see Table 5). They mainly differ in their auxotrophies, principally relying on a histidinol dehydrogenase deficiency (*his4*), allowing, upon transformation, for the positive selection of recombinant expression vectors. Some of them bear additional deficiencies in endogenous proteases (SMD series), while others were recently engineered to perform "human-like" N-glycosylation (26).

As already mentioned in the first section, most of the transforming expression vectors are designed to be maintained as integrative elements in the genome of *P. pastoris*. This is generally achieved through recombination events between linearised sequences borne by the plasmids (typically *HIS4* or P_{AOX1}) and their homologous sequence counterparts present on the genome, leading to the targeted insertion of the expression vectors. Moreover, such plasmid insertions frequently occur in tandem in yeasts and thus lead to multiple integration of the gene of interest with an associated impact on subsequent protein yields.

Table 4
Commercially available expression vectors

Name	Selection marker	Phenotype of transformants	Promoter	Secretion sequence	Added tags
pAO815	*HIS4*	His⁺	P_{AOX1}	None	None
pPIC3.5 K	*HIS4*, Kan	His⁺, G418ᴿ	P_{AOX1}	None	None
pPIC9K	*HIS4*, Kan	His⁺, G418ᴿ	P_{AOX1}	α factor	None
pPICZ A, B, C	Ble	Zeoᴿ	P_{AOX1}	None	c-Myc/his6
pPICZα A, B, C	Ble	Zeoᴿ	P_{AOX1}	α factor	c-Myc/his6
pPIC6 A, B, C	Bsd	Blaᴿ	P_{AOX1}	None	c-Myc/his6
pHIL-D2	*HIS4*	His⁺	P_{AOX1}	None	None
pHIL-S2	*HIS4*	His⁺	P_{AOX1}	PHO1	None
pFLD	Ble	Zeoᴿ	P_{FLD1}	None	V5 epitope/his6
pFLDα	Ble	Zeoᴿ	P_{FLD1}	α factor	V5 epitope/his6
pGAPZ A, B, C	Ble	Zeoᴿ	P_{GAP}	None	c-Myc/his6
pGAPZα A, B, C	Ble	Zeoᴿ	P_{GAP}	α factor	c-Myc/his6
pPink-HC	*ADE2*	Ade⁺	P_{AOX1}	None	None
pPink-LC	*ADE2*	Ade⁺	P_{AOX1}	None	None
pPinkα-HC	*ADE2*	Ade⁺	P_{AOX1}	α factor	None

HIS4 and *ADE2* auxotrophy markers for histidine and adenine, respectively; *Kan*, *Ble* and *Bsd* bacterial genes conferring *P. pastoris* resistance to G418 (or geneticin), zeocin and blasticidin antibiotics, respectively

Alternatively, integration can be obtained by a gene replacement strategy. In this case, a double recombination event must be performed between the *AOX1* promoter and terminator sequences present on the transforming DNA (containing the gene of interest and a selection marker) and the corresponding homologous sequences present within the *P. pastoris* genome. This double recombination event results in the replacement of the *AOX1* gene by the construct of interest.

The phenotype of the resulting transformants is not solely dependent on the selection marker present in the chosen vector (auxotrophy and/or antibiotic resistance). The integration strategy itself dictates the methanol utilisation phenotype of the transformed cells since plasmid insertion results in a Mut⁺ (methanol utilisation plus) phenotype, while the gene replacement of *AOX1* leads to a Mutˢ (methanol utilisation slow) phenotype. In several cases, these differences in methanol utilisation have been reported as an important parameter to consider for enhancing the performance of recombinant protein production (27).

Table 5
***P. pastoris* most commonly used strains**

Strain	Genotype	Phenotype	References
Y11430	Wild-type	Mut⁺	NRRL
X-33	Wild-type	Mut⁺	NRRL
GS115	*his4*	Mut⁺, His⁻	(46)
KM71	*his4, arg4, aox1::ARG4*	Mutˢ, His⁻, Arg⁺	(46)
MC100-3	*his4, arg4, aox1::ARG4, aox2::HIS4*	Mut⁻, His⁺, Arg⁺	(47)
SMD1163	*his4, pep4, prb1*	Mut⁺, His⁻, Prot⁻ (A⁻, B⁻, CarbY⁻)	(48)
SMD1165	*his4, prb1*	Mut⁺, His⁻, Prot⁻ (B⁻)	(48)
SMD1168	*his4, ura3, pep4::URA3*	Mut⁺, His⁻, Prot⁻ (A⁻, Bˢ, CarbY⁻)	(48)
PichiaPink® Strain 1	*ade2*	Mut⁺, Ade⁻,	Invitrogen
PichiaPink® Strain 2	*ade2, pep4*	Mut⁺, Ade⁻, Prot⁻ (A⁻, Bˢ, CarbY⁻)	Invitrogen
PichiaPink® Strain 3	*ade2, prb1*	Mut⁺, Ade⁻, Prot⁻ (B⁻)	Invitrogen
PichiaPink® Strain 4	*ade2, pep4, prb1*	Mut⁺, Ade⁻, Prot⁻ (A⁻, B⁻, CarbY⁻)	Invitrogen

NRRL: Northern Regional Research Laboratories, Peoria, IL

1.7. Where Can I Find a Practical Illustration of the Construction and Preparation of an Expression Vector?

The next sections present the material and protocols needed for the cloning of the gene encoding the adenosine A_{2A} receptor (AA2AR), a G protein-coupled receptor (GPCR), into an engineered pPIC9K plasmid (see Table 1). This vector was modified by standard molecular biology procedures to incorporate a Flag-tag, a TEV protease cleavage sequence and a deca-histidine-tag (10His) upstream of the *Bam*HI and *Spe*I cloning sites for insertion of the target gene, as well as a second TEV site and a biotinylation-tag downstream (28). This combination was selected on the basis of previous studies showing enhanced yields of other GPCRs when fused to the α-MF signal sequence (present on pPIC9K) and the biotinylation-tag (16, 17). The Flag and 10His tag were inserted for detection and purification purposes, the TEV cleavage sites were added to allow cleavage of the N- and C-terminally fused sequences following purification. The brief protocols presented here illustrate a very standard way of generating the desired *P. pastoris* expression plasmid as well as preparation prior to yeast transformation.

2. Materials

2.1. Cloning the Gene of Interest into the Expression Vector

1. A cDNA template containing the full-length AA2AR_HUMAN encoding gene.

2. A 30-base-long AA2A-specific forward primer bearing an additional 5′ adapter specifically designed to introduce a *Bam*HI restriction site (fwd. sequence: 5′-GAAGACA**GGATCC**AT GCCCATCATGGGCTCCTCGGTGTACATC-3′), and a similar reverse primer, bearing a 5′ adapter introducing *Spe*I (rev. sequence: 5′-GAAGACA**ACTAGT**GGACACTCCTGCT CATCCTGGGCCAGGGG-3′) (see Note 1).

3. A high-fidelity polymerase, typically the PrimeSTAR (Takara) or the Phusion (Finnzyme) DNA polymerase, and its specific buffer and dNTP mix.

4. Standard restriction enzymes and their related buffers, here *Bam*HI and *Spe*I (Fermentas, Germany).

5. A T4 DNA ligase, here the Rapid DNA ligation kit (Fermentas).

6. *E. coli* competent cells, here the TOP10 chemically competent cells (Invitrogen).

7. Liquid and agar plates of LB media (1% tryptone, 0.5% yeast extract, 1% NaCl) supplemented with 50 µg/mL kanamycin.

8. A robust nucleic acid extraction and purification kit, here the NucleoSpin kit (Macherey-Nagel).

9. Standard equipment, consumables and chemicals for routine molecular biology techniques including PCR amplification of DNA fragments, DNA separation and visualisation, UV spectrophotometry and *E. coli* culturing.

2.2. Preparation of the Expression Vector

1. Liquid LB medium.
2. NucleoSpin Plasmid kit from Macherey-Nagel.
3. Restriction enzyme *Pme*I and its specific buffer (Fermentas).
4. Phenol.
5. 24:1 (v/v) chloroform-isoamyl alcohol.
6. Ice-cold 100% ethanol.
7. Ice-cold 70% ethanol.
8. 3 M sodium acetate pH 4.8.
9. Sterile H_2O.
10. Agarose gels (1%) supplemented with ethidium bromide.

2.3. Transformation of Pichia pastoris

All materials and solutions must be sterile.

1. YPD-rich medium: 1% yeast extract, 2% peptone, 2% dextrose.
2. Agar plates made with YPD-rich medium supplemented with 2% agar.
3. A fresh SMD1163 colony streaked on a YPD plate.
4. 1 M Hepes pH 8.
5. 1 M dithiothreitol (DTT).
6. 1 M cold sorbitol.
7. Sterile cold H_2O.
8. Electroporation instrument and sterile 0.2-cm electroporation cuvettes.
9. MD plates: 1.34% Yeast Nitrogen Base w/o amino acids, 2% dextrose, 4×10^{-5}% biotin.
10. YPD plates supplemented with 0.1 and 0.25 mg/mL geneticin.

3. Methods

3.1. Cloning the AA2AR Gene into the Modified pPIC9K Expression Vector

3.1.1. PCR Amplification and Preparation of the AA2AR Gene

1. Prepare the PCR reaction mix on ice: typically 1–10 ng of the template cDNA, 5 µL each of the 2 µM stock solution of the forward and reverse primers, 10 µL of 5× PCR buffer, 4 µL of a dNTP mixture (2.5 mM each), 1 U of high fidelity PrimeSTAR polymerase, and sterile water to a final volume of 50 µL.
2. Run the PCR reaction in a thermocycler with a standard 30-cycle protocol alternating 15 s at 98°C, 15 s at 55°C and 1 min at 72°C.
3. Pipette 25 µL of the PCR reaction, add 5 µL 6× loading dye and load the mixture on a 1% agarose gel to analyse the amplified product after migration.
4. Extract the desired DNA fragment using the protocol detailed in the NucleoSpin kit (see Note 2).

3.1.2. Preparation of the Plasmid and Ligation with the Insert DNA

1. In individual eppendorf tubes, cut the amplified insert fragment coding for the gene of interest and the pPIC9K vector with the *Bam*HI and *Spe*I enzymes.
2. Load the digestion products on a 1% agarose gel and following separation extract the cut insert and vector fragments separately (see Note 3).
3. Prepare the ligation reaction with a 5:1 ratio of insert: plasmid. Typically, 50–100 mg of linearised plasmid is used. Add 1 U of T4 DNA ligase together with the ligase buffer and make the final volume to 20 µL with sterile water.

3.1.3. Transformation, Selection and Control of E. coli Recombinant Clones

1. Use about 5 µL of the ligation mixture to transform 50 µL of TOP10 chemically competent cells. Incubate on ice for 5–30 min.

2. Heat-shock the cells for 30 s at 42°C and immediately transfer the tubes on ice.

3. Add 250 µL of regeneration medium (typically SOB or SOC medium) and let the cells regenerate for 1 h at 37°C.

4. Spread 100–200 µL of the transformation mixture on pre-warmed LB agar plates supplemented with 50 µg/mL kanamycin and incubate overnight at 37°C.

5. The following day, pick 6–12 colonies and use to inoculate 2 mL LB supplemented with 50 µg/mL kanamycin. Grow the cultures overnight at 37°C in an incubator shaker. The presence of the insert in a particular clone can also be checked using colony PCR.

6. Purify the plasmid DNA of each clone using the plasmid purification kit following the manufacturer's instructions.

7. Perform restriction digest analysis of the plasmids using the *Bam*HI and *Spe*I enzymes to confirm the presence of the insert.

8. Final check the integrity of the insert by DNA sequencing (see Note 4).

9. Store the plasmid containing the correct sequence at –20°C. In addition, prepare a glycerol stock by adding 700 µL of culture containing correct clone to 300 µL 20% glycerol LB medium in cryotubes and storing at –80°C.

3.2. Preparation of the Expression Vector

3.2.1. Amplification and Linearisation of the Expression Vector

1. Inoculate 5–10 mL LB with the *E. coli* clone containing the expression pPIC9K plasmid and incubate at 37°C overnight.

2. Extract and purify the plasmid DNA using the plasmid preparation kit (Macherey-Nagel) according to the manufacturer's instructions.

3. Prepare a restriction digest solution by adding 5–7 µg of purified plasmid to 25 U of *Pme*I, 20 µL of 10× corresponding buffer and sterile water to a final volume of 200 µl. Incubate the reaction for 2 h at 37°C (see Note 5).

3.2.2. Phenol-Chloroform Extraction of the Linearised Plasmid

1. Add 400 µl of phenol:chloroform (1:1) to the 200 µL digestion mixture.

2. Centrifuge 5 min at $18,000 \times g$ and transfer the superior phase to a new tube.

3. Add 400 µl of chloroform and vortex thoroughly.

4. Centrifuge 5 min at $18,000 \times g$ and transfer the superior phase to a new tube.

5. Add 1 mL of 100% ethanol, 50 µl of 3 M sodium acetate and incubate for at least 1 h at −20°C to precipitate the DNA.

6. Centrifuge for 30 min at 4°C at $18,000 \times g$.

7. Wash the pellet with 100 µL of 70% ethanol.

8. Centrifuge for 5 min at 4°C at $18,000 \times g$.

9. Dry the pellet then resuspend in 15 µL sterile H_2O (see Note 6).

10. Check the DNA linearisation by loading 1 µL of the solution on a 1% agarose gel.

3.3. Transformation of P. pastoris

3.3.1. Preparation of Competent Cells (See Note 7)

1. Inoculate 100 mL of YPG medium with a fresh SMD1163 colony and incubate the preculture overnight at 30°C in an incubator shaker.

2. Measure the OD_{600} of the preculture, dilute it in 400 mL YPG medium in order to obtain an OD_{600} of 0.25 and incubate at 30°C.

3. When the culture reaches an OD_{600} of 1 (after approximately 4 h), harvest the cells by centrifugation in sterile tubes at $2,000 \times g$ and 4°C for 5 min.

4. Resuspend the cells in 100 mL YPD, 20 mL of 1 M Hepes pH 8 and 2.5 mL of 1 M DTT; mix gently until the pellet is resuspended.

5. Incubate for 15 min at 30°C.

6. Transfer onto ice and add sterile and cold H_2O to a final volume of 500 mL.

7. Harvest the cells by centrifugation at $2,000 \times g$ and 4°C for 5 min.

8. Wash the cell pellet with 250 mL of sterile and cold H_2O.

9. Harvest the cells by centrifugation at $2,000 \times g$ and 4°C for 5 min.

10. Resuspend the pellet in 20 mL of cold 1 M sorbitol by gently mixing.

11. Harvest the cells by centrifugation at $2,000 \times g$ and 4°C for 5 min.

12. Resuspend the pellet in 500 µl of cold 1 M sorbitol by gently mixing.

3.3.2. Electro transformation

1. Place an electroporation cuvette on ice at least 10–15 min before performing the transformation.

2. Mix 40 µL of competent cells, with 7.5 µL of the linearised DNA in the cuvettes, mix gently with the pipette and incubate for 5 min on ice.

3. Adjust the electroporation settings as follows: 1,500 V, 25 µF, 600 Ω.

4. Place the cuvette in the electroporator chamber and apply the electric pulse.

5. Immediately resuspend the electroporated mixture in 1 mL of cold 1 M sorbitol and transfer cells to a sterile tube.

6. Allow the cells to recover for about 1 h at 30°C then harvest by centrifugation at $2,000 \times g$ for 10 min.

7. Discard the supernatant and resuspend the pellet in 500 μL 1 M sorbitol.

3.3.3. Selection of Recombinant Clones

1. Spread 2×250 μL of electrotransformed cells on two MD plates and incubate for 2–3 days at 30°C.

2. Harvest the His⁺ transformants by adding 1 mL of YPD onto the plates and scrape off all the clones using a sterile scraper.

3. Perform 10× and 100× dilutions and measure the OD_{600} for each dilution.

4. Spread an equivalent of 10^5 cells/plate (OD_{600} of 1 is equivalent to approximately 5×10^7 cells/mL) on YPD plates supplemented with 0.1 or 0.25 mg/mL geneticin.

5. Incubate for 2–3 days at 30°C. Colonies should only grow if the cells contain the insert together with the selection marker integrated into the genome. Those colonies which grow on plates with the higher geneticin concentration should contain more copies of the selection marker and more copies of the target gene.

4. Notes

1. In the present example, the primers are designed to amplify a gene that will be fused in frame with tag sequences at both 5′ and 3′ ends. Therefore, no stop codon is introduced in the sequence of the reverse primer.

2. If the PCR products appear as pure and single bands corresponding to the desired DNA fragment, you can directly use the PCR mixture in the next step. However, if the PCR reaction results in a number of non-specific products, optimise the reaction conditions by increasing the annealing temperature or the length of the annealing step.

3. If you use restriction enzymes that can be heat inactivated, this step is optional.

4. The sequencing step is mandatory here as the DNA insert has been obtained by PCR. This step is of course not necessary if the insert fragment is obtained from digestion of a previously validated plasmid.

5. Check for the absence of the *Pme*I site in the gene to be expressed, otherwise you will generate several fragments instead of a linearised vector. In case *Pme*I is present, select another enzyme that will cut only once in the vector (*Sac*I or *Sal*I for instance).

6. Make sure the DNA pellet is completely dried, otherwise it will be poorly resuspended and not at the concentration required for transformation.

7. This protocol describes the preparation of electrocompetent cells. Alternative protocols for chemically competent cells are also available in the Invitrogen *Pichia* manual, which is freely available at http://www.invitrogen.com.

References

1. Cereghino JL, Cregg JM (2000) Heterologous protein expression in the methylotrophic yeast *Pichia pastoris*. FEMS Microbiol Rev 24:45–66

2. Cereghino GPL, Cereghino JL, Ilgen C, Cregg JM (2002) Production of recombinant proteins in fermenter cultures of the yeast *Pichia pastoris*. Curr Opin Biotechnol 13:329–332

3. Macauley-Patrick S, Fazenda ML, McNeil B, Harvey LM (2005) Heterologous protein production using the *Pichia pastoris* expression system. Yeast 22:249–270

4. Alkhalfioui F, Logez C, Bornert O, Wagner R (2011) Expression systems: *Pichia pastoris*. In: Production of membrane proteins–strategies for expression and isolation (ed A.S. Robinson) Wiley-VCH Verlag GmbH & Co, DOI: 10.1002/9783527634521

5. Cregg JM, Barringer KJ, Hessler AY, Madden KR (1985) *Pichia pastoris* as a host system for transformations. Mol Cell Biol 5:3376–3385

6. Lee CC, Williams TG, Wong DWS, Robertson GH (2005) An episomal expression vector for screening mutant gene libraries in *Pichia pastoris*. Plasmid 54:80–85

7. Hong IP, Lee SJ, Kim YS, Choi SG (2007) Recombinant expression of human cathelicidin (hCAP18/LL-37) in *Pichia pastoris*. Biotechnol Lett 29:73–78

8. Sandstrom AG, Engstrom K, Nyhlen J, Kasrayan A, Backvall JE (2009) Directed evolution of *Candida antarctica* lipase A using an episomal replicating yeast plasmid. Protein Eng Des Sel 22:413–420

9. Zhang AL, Luo JX, Zhang TY, Pan YW, Tan YH, Fu CY, Tu FZ (2009) Recent advances on the GAP promoter derived expression system of *Pichia pastoris*. Mol Biol Rep 36:1611–1619

10. Hartner FS, Glieder A (2006) Regulation of methanol utilisation pathway genes in yeasts. Microb Cell Fact 5:39

11. Hartner FS, Ruth C, Langenegger D, Johnson SN, Hyka P, Lin-Cereghino GP, Lin-Cereghino J, Kovar K, Cregg JM, Glieder A (2008) Promoter library designed for fine-tuned gene expression in *Pichia pastoris*. Nucleic Acids Res 36:e76

12. Cos O, Ramón R, Montesinos JL, Valero F (2006) Operational strategies, monitoring and control of heterologous protein production in the methylotrophic yeast *Pichia pastoris* under different promoters: a review. Microb Cell Fact 5:17

13. Fantoni A, Bill RM, Gustafsson L, Hedfalk K (2007) Improved yields of full-length functional human FGF1 can be achieved using the methylotrophic yeast *Pichia pastoris*. Protein Expr Purif 52:31–39

14. Delroisse JM, Dannau M, Gilsoul JJ, El Mejdoub T, Destain J, Portetelle D, Thonart P, Haubruge E, Vandenbol M (2005) Expression of a synthetic gene encoding a *Tribolium castaneum* carboxylesterase in *Pichia pastoris*. Protein Expr Purif 42:286–294

15. Daly R, Hearn MTW (2005) Expression of heterologous proteins in *Pichia pastoris*: a useful experimental tool in protein engineering and production. J Mol Recognit 18:119–138

16. Weiss HM, Haase W, Michel H, Reilander H (1995) Expression of functional mouse 5-HT$_{5A}$ serotonin receptor in the methylotrophic yeast *Pichia pastoris*: pharmacological characterization and localization. FEBS Lett 377:451–456

17. Grunewald S, Haase W, Molsberger E, Michel H, Reilander H (2004) Production of the human D2S receptor in the methylotrophic yeast *P. pastoris*. Receptors Channels 10:37–50

18. Fischer G, Kosinska-Eriksson U, Aponte-Santamaria C, Palmgren M, Geijer C, Hedfalk K, Hohmann S, de Groot BL, Neutze R, Lindkvist-Petersson K (2009) Crystal structure of a yeast aquaporin at 1.15 Å reveals a novel gating mechanism. PLoS Biol 7:e1000130

19. Tornroth-Horsefield S, Wang Y, Hedfalk K, Johanson U, Karlsson M, Tajkhorshid E, Neutze R, Kjellbom P (2006) Structural mechanism of plant aquaporin gating. Nature 439:688–694

20. Terpe K (2003) Overview of tag protein fusions: from molecular and biochemical fundamentals to commercial systems. Appl Microbiol Biotechnol 60:523–533

21. Woo JH, Liu YY, Mathias A, Stavrou S, Wang Z, Thompson J, Neville DM Jr (2002) Gene optimization is necessary to express a bivalent anti-human anti-T cell immunotoxin in *Pichia pastoris*. Protein Expr Purif 25:270–282

22. Sinclair G, Choy FY (2002) Synonymous codon usage bias and the expression of human glucocerebrosidase in the methylotrophic yeast *Pichia pastoris*. Protein Expr Purif 26:96–105

23. Tull D, Gottschalk TE, Svendsen I, Kramhøft B, Phillipson BA, Bisgård-Frantzen H, Olsen O, Svensson B (2001) Extensive N-glycosylation reduces the thermal stability of a recombinant alkalophilic bacillus alpha-amylase produced in *Pichia pastoris*. Protein Expr Purif 21:13–23

24. Boettner M, Steffens C, von Mering C, Bork P, Stahl U, Lang C (2007) Sequence-based factors influencing the expression of heterologous genes in the yeast *Pichia pastoris* – a comparative view on 79 human genes. J Biotechnol 130:1–10

25. De Schutter K, Lin YC, Tiels P, Van Hecke A, Glinka S, Weber-Lehmann J, Rouze P, Van de Peer Y, Callewaert N (2009) Genome sequence of the recombinant protein production host *Pichia pastoris*. Nat Biotechnol 27:561–569

26. Hamilton SR, Gerngross TU (2007) Glycosylation engineering in yeast: the advent of fully humanized yeast. Curr Opin Biotechnol 18:387–392

27. Pla IA, Damasceno LM, Vannelli T, Ritter G, Batt CA, Shuler ML (2006) Evaluation of Mut$^+$ and MutS *Pichia pastoris* phenotypes for high level extracellular scFv expression under feedback control of the methanol concentration. Biotechnol Prog 22:881–888

28. André N, Cherouati N, Prual C, Steffan T, Zeder-Lutz G, Magnin T, Pattus F, Michel H, Wagner R, Reinhart C (2006) Enhancing functional production of G protein-coupled receptors in *Pichia pastoris* to levels required for structural studies via a single expression screen. Protein Sci 15:1115–1126

29. Ellis SB, Brust PF, Koutz PJ, Waters AF, Harpold MM, Gingeras TR (1985) Isolation of alcohol oxidase and two other methanol regulatable genes from the yeast *Pichia pastoris*. Mol Cell Biol 5:1111–1121

30. Kobayashi K, Kuwae S, Ohya T, Ohda T, Ohyama M, Tomomitsu K (2000) Addition of oleic acid increases expression of recombinant human serum albumin by the AOX2 promoter in *Pichia pastoris*. J Biosci Bioeng 89:479–484

31. Waterham HR, Digan ME, Koutz PJ, Lair SV, Cregg JM (1997) Isolation of the *Pichia pastoris* glyceraldehyde-3-phosphate dehydrogenase gene and regulation and use of its promoter. Gene 186:37–44

32. Shen SG, Sulter G, Jeffries TW, Cregg JM (1998) A strong nitrogen source-regulated promoter for controlled expression of foreign genes in the yeast *Pichia pastoris*. Gene 216:93–102

33. Tschopp JF, Brust PF, Cregg JM, Stillman CA, Gingeras TR (1987) Expression of the lacZ gene from two methanol regulated promoters in *Pichia pastoris*. Nucleic Acids Res 15:3859–3876

34. Menendez J, Valdes I, Cabrera N (2003) The *ICL1* gene of *Pichia pastoris*, transcriptional regulation and use of its promoter. Yeast 20:1097–1108

35. Sears IB, O'Connor J, Rossanese OW, Glick BS (1998) A versatile set of vectors for constitutive and regulated gene expression in *Pichia pastoris*. Yeast 14:783–790

36. Ahn J, Hong J, Lee H, Park M, Lee E, Kim C, Choi E, Jung J, Lee H (2007) Translation elongation factor 1-alpha gene from *Pichia pastoris*: molecular cloning, sequence, and use of its promoter. Appl Microbiol Biotechnol 74:601–608

37. Ahn J, Hong J, Park M, Lee H, Lee E, Kim C, Lee J, Choi E, Jung J, Lee H (2009) Phosphate-responsive promoter of a *Pichia pastoris* sodium phosphate symporter. Appl Environ Microbiol 75:3528–3534

38. Liu H, Tan XQ, Russell KA, Veenhuis M, Cregg JM (1995) PER3, a gene required for peroxisome biogenesis in *Pichia pastoris*, encodes a peroxisomal membrane-protein involved in protein import. J Biol Chem 270:10940–10951

39. Hashimoto Y, Koyabu N, Imoto T (1998) Effects of signal sequences on the secretion of hen lysozyme by yeast, construction of four secretion cassette vectors. Protein Eng 11:75–77

40. Payne WE, Gannon PM, Kaiser CA (1995) An inducible acid phosphatase from the yeast *Pichia pastoris* – characterisation of the gene and its product. Gene 163:19–26

41. Paifer E, Margolles E, Cremata J, Montesino R, Herrera L, Delgado JM (1994) Efficient expression and secretion of recombinant alpha amylase in *Pichia pastoris* using two different signal sequences. Yeast 10:1415–1419

42. Raemaekers RJM, de Muro L, Gatehouse JA, Fordham-Skelton AP (1999) Functional phytohemagglutinin (PHA) and *Galanthus nivalis* agglutinin (GNA) expressed in *Pichia pastoris* – correct N-terminal processing and secretion of heterologous proteins expressed using the PHA-E signal peptide. Eur J Biochem 265:394–403

43. Kato S, Ishibashi M, Tatsuda D, Tokunaga H, Tokunaga M (2001) Efficient expression, purification and characterization of mouse salivary alpha-amylase secreted from methylotrophic yeast *Pichia pastoris*. Yeast 18:643–655

44. Tokunaga M, Kawamura A, Omor A, Hishinuma F (1992) Structure of yeast pGKL 128-kDa killer-toxin secretion signal sequence: processing of the 128-kDa killer-toxin-secretion-signal –

α-amylase fusion protein. Eur J Biochem 203:415–423

45. Oka C, Tanaka M, Muraki M, Harata K, Suzuki K, Jigami Y (1999) Human lysozyme secretion increased by alpha-factor pro-sequence in *Pichia pastoris*. Biosci Biotechnol Biochem 63:1977–1983

46. Cregg JM, Madden KR (1987) Development of yeast transformation systems and construction of methanol-utilization-defective mutants of *Pichia pastoris* gene disruption. Biol Res Yeasts, II, pp 1–18

47. Cregg JM, Madden KR (1989) Use of site-specific recombination to regenerate selectable markers. Mol Gen Genet 219:320–323

48. White CE, Hunter MJ, Meininger DP, White LR, Komives EA (1995) Large-scale expression, purification and characterization of small fragments of thrombomodulin: the roles of the sixth domain and of methionine 388. Protein Eng 8:1177–1187

Chapter 4

Preparation of *Saccharomyces cerevisiae* Expression Plasmids

David Drew and Hyun Kim

Abstract

Expression plasmids for *Saccharomyces cerevisiae* offer a wide choice of vector copy number, promoters of varying strength and selection markers. These expression plasmids are usually shuttle vectors that can be propagated both in yeast and bacteria, making them useful in gene cloning. For heterologous production of membrane proteins, we used the green fluorescent protein (GFP) fusion technology which was previously developed in the *Escherichia coli* system. We designed an expression plasmid carrying an inducible *GAL1* promoter, a gene encoding a membrane protein of interest and the GFP-octa-histidine sequence. Here we describe construction of multi-copy yeast expression plasmids by homologous recombination in *S. cerevisiae*.

Key words: Green fluorescent protein, Galactokinase, Fluorescence

Abbreviations

GAL	Galactokinase
GPD	Glyceraldehyde-3-phosphate dehydrogenase
TEF	Translation elongation factor 1α
TEV site	Tobacco etch virus protease cleavage site
yEGFP	Yeast-enhanced green fluorescent protein

1. Introduction

One of the advantages working with *S. cerevisiae* is that there is a wide selection of vectors that can be used for various purposes. These vectors were developed and modified during the 1990s and have been continuously optimised for a range of applications.

Roslyn M. Bill (ed.), *Recombinant Protein Production in Yeast: Methods and Protocols*, Methods in Molecular Biology, vol. 866, DOI 10.1007/978-1-61779-770-5_4, © Springer Science+Business Media, LLC 2012

Usually they are shuttle vectors that can be transferred between yeast and bacteria, making them useful tools for gene cloning and molecular biology in general.

Yeast vectors are roughly divided into two groups: one with a yeast centromere sequence (*CEN*) and the other with a 2μ origin of replication. The former is a mitotically stable yeast replicating plasmid with only a single copy present per cell (1); whereas 2μ plasmids have a copy number of about 20 per cell (2). 2μ plasmids are generally more useful for heterologous protein production.

For regulated or constitutive production of heterologous proteins, Mumberg et al. tested various promoters in both the *CEN* and the 2μ plasmids (3, 4). They showed that the galactokinase (*GAL1*) promoter rendered the highest gene expression for controlled and inducible expression of heterologous genes (3). Among the tested constitutive promoters, promoters derived from genes of glyceraldehyde-3-phosphate dehydrogenase (*GPD*) and translation elongation factor 1α (*TEF*) exhibited high strength in heterologous gene expression (4). For both cases, yeast cells carrying vectors of the pRS420 series (2μ plasmids) showed higher protein yields than those containing the pRS410 series (*CEN* plasmids). To further optimise these vectors, they were constructed with four different selection markers, *HIS3*, *TRP1*, *LEU2* or *URA3* (3, 4). Thus, they offer a choice of copy number, promoter and selection marker, so that researchers can choose an expression plasmid according to their needs. A complete set of these expression vectors can be found and purchased from the American Type Culture Collection (ATCC, Rockville, MD, USA).

The heterologous production of membrane proteins has been a challenging task as membrane proteins are often produced in low yields and/or identification of high-yielding systems is time-consuming. To facilitate identification of high-yielding conditions, we took advantage of the green fluorescent protein (GFP) fusion approach that was previously used in *Escherichia coli* (5). It was shown that GFP was fluorescent only when GFP-tagged membrane proteins were correctly targeted to the plasma membrane. If they were aggregated in the cytosol, there was no fluorescent signal. The fluorescence resulting from recombinant membrane protein–GFP fusions could be easily measured both in liquid culture and in standard SDS gels.

We designed an expression vector carrying either constitutive, *TEF*, or regulated *GAL1* promoters in a 2μ plasmid (pRS426) with C-terminal GFP and an octa-histidine tag for detection and purification of proteins (6). A yeast codon optimised Tobacco Etch Virus (TEV) protease site was designed upstream of GFP for removal of the GFP-His$_8$ tag.

When we initially tested production levels of more than 20 membrane proteins using the two vectors above in *S. cerevisiae*, we found that the *GAL1* promoter gave overall higher expression than the constitutive *TEF* promoter (6). Further, we found that membrane yields were enhanced in the *pep4* deletion strain in which the vacuolar endopeptidase Pep4p is deleted. Deletion of the *pep4* gene not only inhibits Pep4p protease activity but also reduces the levels of other vacuolar hydrolases (7).

Here we describe the practical steps in using homologous recombination to generate sequences encoding a membrane protein–GFP-His$_8$ tag fusion in the 2μ yeast vector carrying a *GAL1* promoter, pRS426GAL1. This strategy has been successfully employed by others to overproduce eukaryotic membrane proteins (8).

2. Materials

1. Yeast *pep4* deletion strain, FGY217 (*MATa, ura3-52, lys2Δ201, pep4Δ*) (9).

2. A vector containing the yeast-enhanced green fluorescent protein (pUG35) (10).

3. Growth medium without uracil (for 1 L, 6.7 g yeast nitrogen base without amino acids (BD Difco, cat. No. 291920)), 2 g yeast synthetic drop-out medium supplement without uracil (Sigma, cat. No. Y1501), and either 2% glucose (for pre-culture) or 0.1% glucose (for expression culture). For plates, add 20 g of bacto agar (Sigma, cat No. A5306). D-(+) glucose can be purchased from Sigma (cat. No. G7021).

4. 20% galactose (w/v) (Sigma, cat. No. G0625).

5. *Sma*I restriction enzyme (Invitrogen, cat. No. 15228018).

6. For yeast transformation with lithium acetate, YPD medium (BD Difco, cat. No. 242810), polyethylene glycol (PEG) 3350 (Sigma, cat. No. P3640), lithium acetate (Sigma, cat. No. L4158), salmon sperm DNA (Sigma, cat. No. D1626), dimethyl sulfoxide (DMSO) (Sigma, cat. No. D2438) are needed.

7. PCR reagents (polymerases and dNTPs are available from various companies).

8. Mini-prep plasmid kit (Qiagen).

9. Acid-washed glass beads, 500 μm (Sigma).

3. Methods

3.1. Construction of an Expression Vector Carrying the TEV Site-GFP-His$_8$ Sequence

1. Amplify a fragment containing the yeast-enhanced GFP sequence (yEGFP) excluding the start codon from the pUG35 plasmid (10) by PCR using two primers: forward 5′-TCTAGAACTAGTGGATCCCCCCCCGGGGAAAATT TATATTTTCAAGGTC-3′ and reverse 5′-GACGGTATCGAT AAGCTTGATATCAATTCCTGCAGTTAATGATGATG-3′.

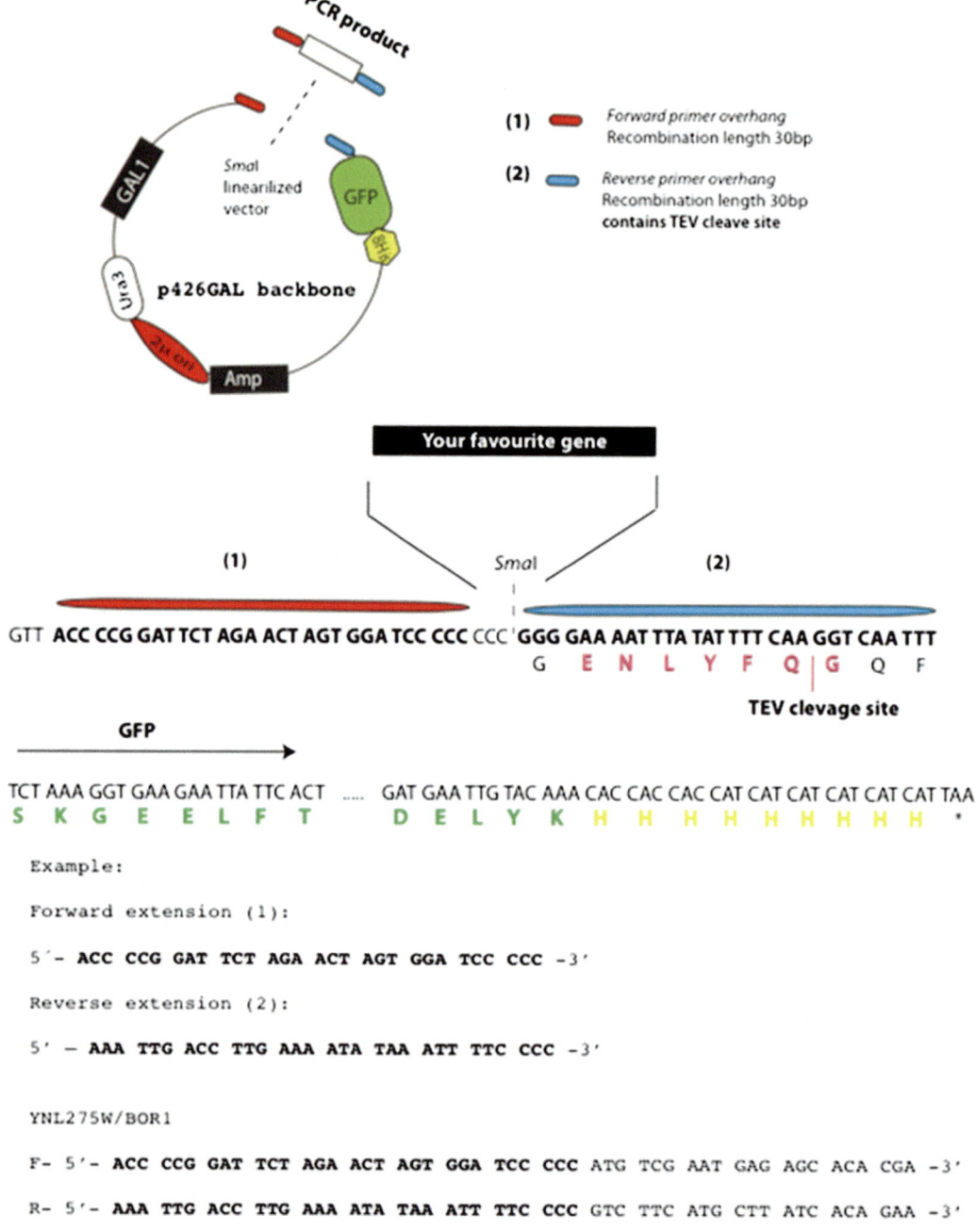

Fig. 1. Cloning by homologous recombination into a 2µ *S. cerevisiae* GFP-fusion expression vector (11). Primer design for the gene YNL275W/BOR1 is shown as an example. The sequences in *bold* are the homologous recombination sites to the vector.

These primers contain sequences encoding (1) tobacco etch virus (TEV) cleavage sequence (ENLYFQG/S), (2) *SmaI* endonuclease restriction site, (3) octa-histidine tag on the reverse primer and (4) homologous recombination sites to the pRS426 plasmids (Fig. 1) (11).

2. Digest pRS426GAL1 with a *SmaI* restriction enzyme to linearise the vector (see Note 1).

3. Transform any uracil auxotrophic *S. cerevisiae* strain with the PCR-amplified fragment and the *SmaI*-digested pRS426GAL1 plasmid using lithium acetate (12). Select on medium lacking uracil. To assess the number of background colonies carrying only an empty vector, prepare one sample lacking the PCR product. The number of these background colonies should be very low.

4. Inoculate yeast transformants in 5 mL of medium lacking uracil, grow overnight at 30°C and isolate plasmids by following the mini-prep plasmid kit manufacturer's protocol with the exception that the harvested cells should be suspended in the kit's resuspension buffer (the first buffer in the mini-prep kit). Then, add an equal volume of acid-washed glass beads to the cell suspension, vortex at maximum speed for 5 min, collect the glass beads and unbroken cells with a brief centrifugation step and transfer the supernatant to a new tube. Thereafter proceed to the next step of the mini-prep kit protocol. These plasmids should be further amplified and isolated from bacterial cells for DNA sequencing. The pRS420 series of vectors contains an ampicillin-resistance marker for selection in bacterial cultures.

5. Isolate plasmids from bacterial cells and sequence to confirm an in-frame TEV site-GFP-His$_8$ sequence. pRS426GAL1 carrying the TEV-site-GFP-His$_8$ sequence is designated pDDGFP-2 (6).

3.2. Cloning of Membrane Protein-Encoding Gene(s) into the pDDGFP-2 Plasmid

1. Amplify the gene of interest with primers that include 5′ overhangs (35 bp) complementing the upstream and downstream sequences to either side of the *SmaI* site in the pDDGFP-2 vector (Fig. 1) (11).

2. Carry out transformation of FGY217 (*MATa, ura3-52, lys2Δ201, pep4Δ*) with the PCR-amplified gene and *SmaI*-linearised pDDGFP-2 (9) using lithium acetate (11, 12) and select on medium lacking uracil. As for the pDDGFP-2 vector above, prepare one sample without the PCR product to assess the number of background colonies. Yeast plates can be stored at 4°C for at least 1 month without a reduction in production levels.

3. To confirm that a colony harbours the vector and gene of interest, inoculate 5 mL of medium lacking uracil with the colony, grow overnight at 30°C and either do yeast colony PCR using sequencing primers or check the fluorescence of whole cell lysates (described in detail in Chapter 8).

4. Note

1. Digestion with *Sma*I gives rise to blunt ends that eliminate background colonies caused by re-annealing of a linearised vector by homologous recombination during transformation. There may be a higher number of background colonies when using restriction enzymes that produce sticky ends. This does not appear to be reduced by including a dephosphorylation step.

Acknowledgments

This work was supported by the Royal Society (United Kingdom) through a University Research Fellowship to DD and by a Basic Science Research Program grant through the National Research Foundation of Korea (NRF) funded by the Ministry of Education, Science and Technology (NRF0409-20100093) to HK.

References

1. Sikorski RS, Hieter P (1989) A system of shuttle vectors and yeast host strains designed for efficient manipulation of DNA in *Saccharomyces cerevisiae*. Genetics 122:19–27

2. Christianson TW, Sikorski RS, Dante M, Shero JH, Hieter P (1992) Multifunctional yeast high-copy-number shuttle vectors. Gene 110:119–122

3. Mumberg D, Muller R, Funk M (1994) Regulatable promoters of *Saccharomyces cerevisiae* – comparison of transcriptional activity and their use for heterologous expression. Nucleic Acids Res 22:5767–5768

4. Mumberg D, Muller R, Funk M (1995) Yeast vectors for the controlled expression of heterologous proteins in different genetic backgrounds. Gene 156:119–122

5. Drew DE, von Heijne G, Nordlund P, de Gier JW (2001) Green fluorescent protein as an indicator to monitor membrane protein overexpression in *Escherichia coli*. FEBS Lett 507:220–224

6. Newstead S, Kim H, von Heijne G, Iwata S, Drew D (2007) High-throughput fluorescent-based optimization of eukaryotic membrane protein overexpression and purification in *Saccharomyces cerevisiae*. Proc Natl Acad Sci USA 104:13936–13941

7. Woolford CA, Daniels LB, Park FJ, Jones EW, Van Arsdell JN, Innis MA (1986) The *PEP4* gene encodes an aspartyl protease implicated in the posttranslational regulation of *Saccharomyces cerevisiae* vacuolar hydrolases. Mol Cell Biol 6:2500–2510

8. Li M, Hays FA, Roe-Zurz Z, Vuong L, Kelly L, Ho CM, Robbins RM, Pieper U, O'Connell JD 3rd, Miercke LJ, Giacomini KM, Sali A, Stroud RM (2009) Selecting optimum eukaryotic integral membrane proteins for structure determination by rapid expression and solubilization screening. J Mol Biol 385:820–830

9. Kota J, Gilstring CF, Ljungdahl PO (2007) Membrane chaperone Shr3 assists in folding amino acid permeases preventing precocious ERAD. J Cell Biol 176:617–628

10. Cormack BP, Bertram G, Egerton M, Gow NA, Falkow S, Brown AJ (1997) Yeast-enhanced green fluorescent protein (yEGFP): a reporter of gene expression in *Candida albicans*. Microbiology 143:303–311

11. Drew D, Newstead S, Sonoda Y, Kim H, von Heijne G, Iwata S (2008) GFP-based optimization scheme for the overexpression and purification of eukaryotic membrane proteins in *Saccharomyces cerevisiae*. Nat Protoc 3:784–798

12. Amberg DC, Burke DJ, Strathern JN (2005) Methods in yeast genetics: a cold spring harbor laboratory course manual. Cold Spring Harbor Laboratory Press, New York, USA.

Chapter 5

Codon Optimisation for Heterologous Gene Expression in Yeast

Kristina Hedfalk

Abstract

Heterologous protein production is used to amplify the yield of a desired protein target. To date, however, this is not a streamlined process: the factors defining an optimal protein production experiment are still poorly understood. This empirical exercise is particularly challenging for proteins of eukaryotic origin as well as those located in cellular membranes. The strong interest in structural and functional characterisation of eukaryotic membrane proteins—of which many are targets for different drugs—means that large amounts of pure protein, and hence high production levels in a suitable host, are required. On the genetic level, there are mainly two ways to positively influence the final yield of a desired protein target. First, the sequence surrounding the starting ATG can be altered and second the genetic code itself can be optimised to suit the selected host for production. The practical aspects of these two strategies will be discussed and exemplified in further detail in this chapter together with some hints and troubleshooting around different stages of the procedure.

Key words: Protein production, Codon optimisation, Kozak consensus sequence, PCR, Primers

1. Introduction

Protein production in any organism is to some extent a trial and error exercise where the general principle is that eukaryotic proteins are more difficult to produce compared to their prokaryotic counterparts and membrane-bound proteins are less straightforward to produce compared to soluble proteins. While production of bacterial proteins in the historically most well-established host for protein production, namely *Escherichia coli*, can nearly be regarded as a routine strategy, production of, for example, human proteins is much less standardised. However, production of eukaryotic proteins is generally more successful in a eukaryotic host compared to a prokaryotic host. Hence, yeast provides an attractive host for

Roslyn M. Bill (ed.), *Recombinant Protein Production in Yeast: Methods and Protocols*, Methods in Molecular Biology, vol. 866, DOI 10.1007/978-1-61779-770-5_5, © Springer Science+Business Media, LLC 2012

production of eukaryotic proteins including those of human origin. While being a eukaryotic cell, yeast is a simple organism that is easy to grow and manipulate, and in that sense comparable to its prokaryotic counterpart, *E. coli*. Besides this, the underlying limitations to efficient protein production are not well understood, making the troubleshooting and optimisation of the production more difficult. In recent times, however, some knowledge of the optimisation of recombinant proteins has started to accumulate adding more opportunities for designing a high-yielding protein production experiment. This chapter will describe what can be done at the genetic level in order to positively influence the final protein yield: relatively small changes in the nucleotide sequence can have a big impact on protein production.

Besides the fact that there is a reasonable chance of success when choosing yeast for production of a eukaryotic protein, there are, in principle, two reasons to optimise the gene in order to increase the final protein yield: optimise the consensus sequence around the starting ATG to ensure efficient translation initiation as well as optimise the codons to be host specific. The different aspects of these two strategies will be described in further detail below as well as examples thereof.

1.1. Efficient Translation Initiation

The sequence surrounding the start methionine has been shown to influence translation initiation (1), hence this sequence is important for efficient protein production. For mammalian proteins, the Kozak consensus sequence has been determined to be gccRccatgG, where R is a purine (adenine or guanine) and most often an adenine (2). Part of the underlying reason for this specific sequence is not related to translation initiation, rather the +4 G gives rise to an overrepresentation of alanine (GCN) and to a smaller extent glycine (GGN) at the second codon (3). According to this proposed "amino acid constraint hypothesis", these small aliphatic amino acids are needed as the second amino acid for efficient cleavage of the initiator methionine. In comparison, a translation initiator sequence has been presented in yeast, AAAATGTCT, resulting in a serine at the second position but still having the adenine nucleotide preference at the −3 position (4). The absolute preference for the ideal sequence surrounding the starting ATG is not fully clear, and each of these sequences has alternated as recommended in, for example, different versions of the "Invitrogen Easy Select *Pichia* Expression Kit" manual (5). Some light will, however, be shed on which sequence should be used in the example below (also see Note 1).

1.1.1. The Influence of Nucleotides Flanking the Start Codon; Production of 13 Human Aquaporins in Pichia pastoris

In order to get an increased understanding of the bottleneck to a successful eukaryotic membrane protein production experiment, a set of homologous proteins was produced in a suitable host. More precisely, 13 human aquaporins (AQP) were produced in the methylotrophic yeast *P. pastoris* (6). The overall conclusion from

Table 1
Relative production levels of various AQPs

	Second codon	Second amino acid	Relative production level[a]
hAQP8 *wt*	TCT	Serine	0.005
hAQP8 *mut*	GCT	Alanine	0.015
hAQP1 *wt*	GCC	Alanine	1.5
hAQP1 *mut*	GCT	Alanine	0.9

[a]The total production is related to the production of SoPIP2;1, the water channel from spinach, which is produced to high levels in *P. pastoris* membranes, extracted, purified and crystallised to high resolution (7)

this study is that small changes in the nucleotide or protein sequence could influence the maturation, folding and stability of the nascent peptide chain when produced in yeast, which in turn has a major impact on the final protein yield. Noteworthy, the importance of the precise sequence of the triplet for the second codon was confirmed: a G in the +4 position was dominant among the well-producing AQPs while thymine was common among moderate to low producing AQPs, supporting a beneficial use of the mammalian Kozak sequence (2) over the yeast consensus sequence (4) when producing proteins in *P. pastoris*. For example, hAQP8 is produced to relatively low levels in *P. pastoris*. However, mutating the hAQP8 wild-type sequence at the second codon, for which the endogenous sequence is identical to the yeast consensus sequence, to contain a G in the +4 position increased the production about three times (Table 1) supporting the notion that a change from serine to alanine in this position is beneficial for protein production. In comparison, keeping the wild-type, Kozak-like sequence for hAQP1 was beneficial compared to mutating it to mimic the yeast consensus sequence only allowing silent mutations and, hence, keeping the alanine as the second amino acid. Taken together, these two observations support an interpretation that higher protein production is more likely to be achieved in yeast when using the mammalian Kozak sequence instead of the yeast consensus sequence (see Note 2).

1.2. Codon Optimisation

With 64 possible codons but only 20 amino acids, the genetic code is clearly degenerated giving rise to more than one possible codon for each amino acid. As a result, the majority of amino acids, methionine and tryptophan being the only exceptions, have more than one triplet coding for them. When there are alternatives, each organism has its own preferred codon for each individual amino acid. Hence, more efficient translation is likely when adapting a

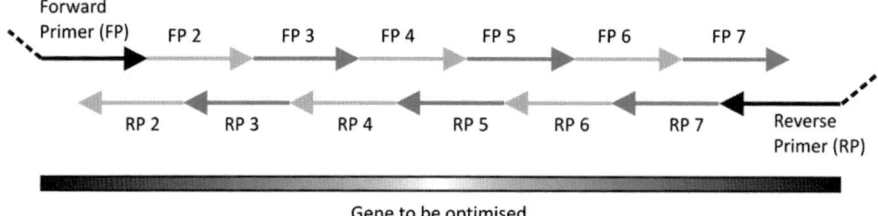

Fig. 1. Amplification of a full-length gene using multiple, codon-optimised primers.

gene sequence so that it contains the preferred codons of the host organism; i.e. to express the gene of interest from an optimised synthetic gene coding for the protein of interest (see Note 3). The frequency of the occurrence of each triplet in a certain organism can be found at http://www.kazusa.or.jp/codon, which provides very useful information for synthetic gene design.

There are two ways to achieve a synthetic gene designed for expression in a certain host; it can either be made by a two step PCR reaction or it can be made commercially by a specialised company. In the first example, the gene is designed selecting the optimal codons for the specific organism. Such a sequence can be made by using The CodonAdaptationToll (JCat) (8), http://www.jcat. de, which adapts the codon usage to most sequenced prokaryotic organisms and a selection of eukaryotic organisms. In addition, using codon adaptation with JCat one can avoid unwanted cleavage sites as well as Rho-independent transcription terminators. For the PCR reaction, overlapping primers are designed (see Note 4) to be flanked by one forward and one reverse primer containing the appropriate restriction sites for the subsequent cloning (see Note 5 and Fig. 1). For a successful first PCR reaction, the primers must not only be codon optimised but also suitable for the reaction as such. Computer programs are available, e.g. DNAWorks (9), that provide a set of oligonucleotides with highly homogenous melting temperatures combined with a minimised tendency for hairpin formation.

The quicker, but certainly more expensive, alternative is to buy the codon-optimised gene from a company, for example Geneart; http://www.geneart.com, which is based on intellectual property in patent WO2004059556 (Method and device for optimising a nucleotide sequence for the purpose of expression of a protein). The method used by this specific company takes sequence repeats, codon usage, GC content, polyA sites, splice sites and RNA secondary structure into account when designing the optimised gene.

1.2.1. Codon Optimisation of a Heterologous Gene Expressed in P. pastoris, as Exemplified by PfAQP

One example of successful production of a codon-optimised gene is the production of the aquaporin from the malaria parasite *Plasmodium falciparum*, PfAQP, in *P. pastoris* (10). PfAQP is a putative drug target for malaria, hence further characterisation of the isolated protein with respect to its function and structure is desired. For this to be achieved, a high production level in a suitable host is needed. *P. falciparum* is a eukaryotic microorganism and yeast, being a eukaryote, was the system that, in addition to the traditional host, *E. coli*, was tested for protein production. In *E. coli*, PfAQP was produced to low levels and the appropriate optimisation of the gene only slightly improved the yield. On the other hand, in *P. pastoris* the wild-type gene was not produced at all which may in part be explained by the difference in GC content: 29% in *P. falciparum* compared to about 50% in *P. pastoris*. On the contrary, the codon-optimised gene (purchased from Geneart) resulted in a high production level of membrane-localised and functional protein in *P. pastoris*. Noteworthy, 72% (189) of the codons were changed in the optimised gene giving rise to a final GC content of 44%. The final yield of PfAQP was estimated to be 18 mg pure protein per litre fermentor culture, which is enough for further functional and structural characterisation.

2. Materials

2.1. Primer Design

For a successful PCR reaction, well-designed primers are essential (see Note 4). In general, there are several considerations when designing a primer for PCR;

1. The length of the primer should be about 20 nucleotides, but the length could be varied in order to achieve the desired properties. However, for gene optimisation using the overlapping primer approach, the primer could be around 50 nucleotides in order to cover the whole gene with fewer primers.

2. The GC content should be around 50% to get a suitable melting temperature of the primer.

3. If possible, the primer should end with a G/C-rich stretch to achieve strong binding in the 3′ end from which the new strand will be built.

4. The primer should not contain any internal repeats or form hairpin loops.

5. The primer should not form strong duplexes with itself.

6. Suitable sequences for restriction digest for subsequent cloning are added on the flanking forward and reverse primer, respectively (see Notes 5 and 6).

7. There are several Web-based tools available for primer analysis, for example http://www.sigma-genosys.com/calc/DNACalc.asp.

2.2. PCR

1. Template containing the gene to be optimised/amplified; 50 ng/μL (see Notes 6 and 7).

2. Forward primer, 10 μM.

3. Reverse primer, 10 μM.

4. Primer cocktail for gene optimisation; 2 μM of each primer. For codon optimisation only.

5. dNTPs, 10 mM.

6. Phusion® Hot Start High-Fidelity DNA Polymerase from Finnzymes, 2 U/μL (see Note 8).

7. 5× Phusion® High-Fidelity buffer (HF buffer) from Finnzymes.

8. TE Buffer; 10 mM Tris–HCl (pH 8.0), 1 mM EDTA (pH 8.0).

9. For the PCR reaction a thermocycler is needed, preferably with a heated lid which prevents condensation of water in the lid of the tube. This means the addition of oil on top of the reaction can be avoided. Ideally, a temperature ramp is programmed over the block in order to easily screen for the optimal annealing temperature for the reaction.

3. Methods

3.1. The PCR Reaction

The PCR (Polymerase Chain Reaction) is a 25-year old method which is used to amplify genetic material and produce large amounts of a certain genetic sequence. The PCR reaction is also a very useful tool to alter the original DNA sequence; introduce mutations or optimise the sequence. For the reaction to proceed, two short oligonucleotides, primers, are needed flanking the region that will be amplified. The reaction takes place in several steps. First, the double stranded template DNA strands are denatured at about 95°C to allow strand separation. In the following step, the solution is cooled down allowing the primers to anneal to the single stranded template strands. During the third step, new strands are synthesised by a heat tolerant DNA-polymerase adding free dNTPs (deoxyribo*N*ucleotide *Tri*Phosphates) to the growing strand. These three steps are repeated 20–30 times giving an exponential amplification of the DNA sequence resulting in 2^n DNA molecules, where n is the number of reaction cycles.

3.1.1. PCR Reaction for Gene Synthesis Using Overlapping Primers (Codon Optimisation)

In order to synthesise a codon-optimised gene, the following mixture is used for the PCR reaction: Primer cocktail, 1 μL (2 pmol/primer); dNTP, 2 μL (20 nmol); 5× HF buffer, 10 μL (1×); DNA Polymerase, 1 μL (2 U); and water, 36 μL, to a final volume of 50 μL.

3.1.2. PCR Reaction for Ordinary Gene Amplification

In order to amplify a gene, the following mixture is used for the PCR reaction: Template, 1 μL (50 ng); forward primer, 5 μL (50 pmol); reverse primer, 5 μL (50 pmol); dNTP, 2 μL (20 nmol); 5× HF buffer, 20 μL (1×); DNA Polymerase, 1 μL (2 U); and water, 66 μL, to a final volume of 100 μL.

3.2. PCR Program

1. 98°C; 30 s
2. 98°C; 10 s
3. 58 ± 10°C; 30 s
4. 72°C; 30 s
5. 72°C; 10 min
6. 4°C; hold

Steps 2–4 are repeated 30 times.

3.3. Procedure

1. Primers are usually delivered in the form of a freeze dried DNA pellet. For long-term storage, a 200-µM stock solution is made in TE Buffer, pH 8.0, which is stored at –20°C.

2. Primer working solutions, 2–10 µM, for the PCR reaction are made in H_2O to avoid high concentrations of EDTA in the resulting reaction.

3. In order to find the optimal annealing temperature for a specific reaction, the reaction mixture is divided into 10–12 tubes and the PCR program is run using a gradient of 20 degrees, 58 ± 10°C, over the PCR block.

4. The whole reaction from the use of each annealing temperature in step 3 above, 8–10 µL, is analyzed on an analytical agarose gel.

5. For codon optimisation using overlapping primers, reagents are mixed according to Subheading 3.1.1 and a 50-µL reaction is run to make the codon-optimised template.

6. The reaction in step 5 above, 1 µL, is used as template for the next PCR reaction and amplification of the gene. Reagents are mixed according to Subheading 3.1.2.

7. As a control for the PCR reaction, one tube lacking template should be run in parallel to verify the specificity of the reaction.

8. The reaction, 5 µL, is analyzed using an analytical agarose gel.

4. Notes

1. Based on the production of 13 human aquaporins in *P. pastoris*, the Kozak sequence should be introduced surrounding the starting ATG.

2. An A in the –3 position is simple to add, but the +4 G is less trivial since it leads to a mutation. Therefore, it could be wise to try two constructs in parallel; one with the G added and one keeping the wild-type sequence at the +4 position.

3. For a novel gene to be produced in yeast it is always worthwhile to try the wild-type sequence in parallel with optimised versions. It could be valuable to compare the codon usage between the gene of interest and yeast in order to determine whether codon optimisation should be prioritised. However, even with very similar Codon Adaptation Indices (CAI) (11), codon optimisation could substantially increase production levels, hence, such an approach could be a valuable option for proteins with an intrinsically low production yield.

4. To make an optimal primer, several aspects should be taken into account (see Subheading 2.1 above). The reality is, however, that most often the main sequence of the primer cannot be chosen freely since the defined location of the primer in the sequence restricts the opportunities to change the nucleotides. Hence, it may be necessary to try primers empirically and then decide on further optimisation.

5. For a successful restriction digest and cloning it is important to have sufficient flanking region at the end of the PCR product to which the specific restriction enzyme can bind. The requirements for the most abundant enzymes can be found at http://www.fermentas.com/en/support/technical-reference/restriction-enzymes/cleavage-efficiency.

6. In the vast majority of cases, the PCR reaction does not cause any problems. However, if no product is obtained, the first thing to check is the quality of the DNA preparation. The next level of troubleshooting is alternative primers still flanking the region of interest but with altered lengths as well as modifications of the added restriction sites or the nonsense sequence in the primer ends.

7. Template for the PCR reaction can be derived from an ordinary plasmid mini preparation and the volume given in Subheading 3.1.2 can be adjusted depending on the resulting concentration from the DNA preparation.

8. The enzyme suggested in Subheading 2.2 is successfully used in the work described here; however, there are many alternative DNA Polymerases for this purpose. If the PCR reaction still does not result in any production despite optimisation trials, another DNA polymerase could be tried.

Acknowledgements

This work was supported by the European Commission (SPINE, MalariaPorin project, E-MeP), the Chalmers Bioscience Initiative and the Wallenberg Foundation (Membrane Protein Center, Lundberg Laboratory, Göteborg), the Swedish Research Council (VR) as well as the Research School of Genomics and Bioinformatics.

References

1. Kozak M (1984) Point mutations close to the AUG initiator codon affect the efficiency of translation of rat preproinsulin *in vivo*. Nature 308:241–246

2. Kozak M (1987) An analysis of 5′-noncoding sequences from 699 vertebrate messenger RNAs. Nucleic Acids Res 15:8125–8148

3. Xia X (2007) The +4 G site in Kozak consensus is not related to the efficiency of translation initiation. PLoS One 2:e188

4. Cigan AM, Donahue TF (1987) Sequence and structural features associated with translational initiator regions in yeast – a review. Gene 59:1–18

5. Invitrogen (2009) EasySelect *Pichia* Expression Kit, version I. http://www.invitrogen.com/content/sfs/manuals/easyselect_man.pdf

6. Oberg F, Ekvall M, Nyblom M, Backmark A, Neutze R, Hedfalk K (2009) Insight into factors directing high production of eukaryotic membrane proteins; production of 13 human AQPs in *Pichia pastoris*. Mol Membr Biol 26:215–227

7. Tornroth-Horsefield S, Wang Y, Hedfalk K, Johanson U, Karlsson M, Tajkhorshid E, Neutze R, Kjellbom P (2006) Structural mechanism of plant aquaporin gating. Nature 439:688–694

8. Grote A, Hiller K, Scheer M, Munch R, Nortemann B, Hempel DC, Jahn D (2005) JCat: a novel tool to adapt codon usage of a target gene to its potential expression host. Nucleic Acids Res 33:W526–W531

9. Hoover DM, Lubkowski J (2002) DNAWorks: an automated method for designing oligonucleotides for PCR-based gene synthesis. Nucleic Acids Res 30:e43

10. Hedfalk K, Pettersson N, Oberg F, Hohmann S, Gordon E (2008) Production, characterization and crystallization of the *Plasmodium falciparum* aquaporin. Protein Expr Purif 59:69–78

11. Sharp PM, Li WH (1987) The codon Adaptation Index – a measure of directional synonymous codon usage bias, and its potential applications. Nucleic Acids Res 15:1281–1295

Chapter 6

Yeast Transformation to Generate High-Yielding Clones

Mohammed Jamshad and Richard A.J. Darby

Abstract

There are several ways to introduce non-native DNA into yeast cells, including chemical transformation and electroporation. Methods for both of these procedures are outlined in this chapter. Both methods permit the uptake of DNA from the environment through yeast cell membranes and this DNA can be episomally maintained or integrated into the host genome. However, yeast cells must first be made competent to permit passive entry of the DNA and various methods are outlined in this chapter to facilitate this. All of the described methods can be applied in combination with antibiotic or auxotrophic selection pressure.

Key words: Electroporation, *Pichia pastoris*, *Saccharomyces cerevisiae*, Competent cells

1. Introduction

As a prelude to using yeast as a host for the production of heterologous proteins the cells must first be transformed, and this is entirely dependent on the availability of high-quality vector DNA harboring the gene of interest. This DNA must be obtained by prior amplification of the vector in a bacterial host such as *Escherichia coli* XL1-Blue (Stratagene), DH5α, or Top10 (Invitrogen). *E. coli* clones containing the yeast expression vector are cultured, lysed, and the vector DNA purified from the lysate using commercially available kits such as QIAprep spin kits.

1.1. Transformation of Pichia pastoris

Having prepared the expression vector the DNA can be integrated into the *AOX1* locus of the *Pichia pastoris* genome by homologous recombination giving rise to an expression strain. This is achieved by restriction digestion of the vector with *Bst*XI, *Pme*I, or *Sac*I which yields a linear product. The choice of restriction enzyme will be governed by the restriction site compatibility of the gene cloned into the expression vector.

Roslyn M. Bill (ed.), *Recombinant Protein Production in Yeast: Methods and Protocols*, Methods in Molecular Biology, vol. 866, DOI 10.1007/978-1-61779-770-5_6, © Springer Science+Business Media, LLC 2012

It is generally recognized that transformation of *P. pastoris* cells is less efficient than that of *Saccharomyces cerevisiae,* and it is for this reason that the authors recommend electroporation as the technique for transforming *P. pastoris* when using Invitrogen's pPICZ series of expression vectors. However, an alternative transformation method using PEG 1000 is also presented for those labs that do not have access to an electroporator. It should be noted, however, that this method has a much lower efficiency than electroporation which yields approximately 300 colonies per μg of vector. Both methods utilize the bleomycin-resistance gene (1–4) to confer resistance to zeocin.

Preparation of electro-competent cells is achieved by washing mid-log phase cells with a buffering solution. This eliminates media components, such as salts, which would result in arcing during electroporation. Many protocols for the preparation of competent cells require multiple washes, centrifugation and incubation steps which result in a time-consuming protocol. We recommend a condensed protocol with fewer manipulations, which yields comparable efficiencies and provides sufficient transformants for selecting multicopy integrants (5).

A "recovery" step is required post electroporation as the cells are fragile and require careful handling prior to exposure to selection pressure for transformants. This recovery is conducted in a rich medium allowing the cells the opportunity to regain membrane integrity and requires a minimum of 45 minutes to achieve this. If the selection method relies on an antibiotic such as zeocin or geneticin (G418), clones possessing multiple insertions may be isolated. These so-called jackpot clones can be encouraged by plating out on multiple agar plates with increasing antibiotic concentrations: clones with multiple copies of the integrated vector will be more tolerant to the antibiotic and may yield a two- to threefold increase in recombinant yields (6).

1.2. Transformation of *Saccharomyces cerevisiae*

Transformation of *S. cerevisiae* cells with episomal DNA vectors can be achieved by a number of methods including lithium acetate (LiOAc) transformation and electroporation. In this case, selective pressure (usually auxotrophic) must be maintained by growing the cells on a nutrient-limited medium in order to retain the expression vector. The LiOAc method discussed below uses carrier DNA (e.g. sonicated salmon sperm DNA) to aid uptake of the expression vector. *S. cerevisiae* cells are cultured in a complex medium (e.g. YPD, yeast extract, peptone, and glucose) to mid/late log phase and then incubated in transformation solution with both the carrier DNA and vector. Following an incubation period, the cells are removed, plated on selective agar, and incubated until colonies appear. The composition of the selective agar is dependent upon the auxotrophic marker present in the vector.

The preparation method for electro-competent *S. cerevisiae* cells discussed here does not permit their long-term storage at −80°C and it is recommended that they are made fresh on the day if possible.

2. Materials

1. Expression vector, *store at −20°C.*
2. Carrier DNA: 7 mg/mL sonicated salmon sperm DNA, *store at −20°C.*
3. Transformation Solution: polyethylene glycol 3350 40% (w/v), 1 mM EDTA, 10 mM Tris–HCl pH 7.5, 0.1 M lithium acetate, *stable at room temperature.*
4. YPD medium: 1% yeast extract, 2% bacto peptone, 2% glucose, *stable at room temperature.*
5. BEDS solution: 10 mM bicine-NaOH pH 8.3, 3% (v/v) ethylene glycol, 5% (v/v) (dimethyl sulfoxide) DMSO, 1 M sorbitol, *stable at 4°C for up to 2 months.*
6. PEG 1000 Solution: polyethylene glycol 1000 40% (w/v), 200 mM bicine, pH 8.35.
7. Bicine solution: 10 mM bicine pH 8.35, 150 mM NaCl.
8. YPDS Agar: 1% yeast extract, 2% peptone, 2% dextrose, 1 M sorbitol, 2% agar, *stable at 4°C for up to 2 weeks in the dark.*
9. 1 M dithiothreitol: *store at −20°C.*
10. Eppendorf multiporator™ (Hamburg).
11. Electroporation cuvettes: 2-mm path length.
12. *S. cerevisiae* or *P. pastoris* freshly plated.
13. 10× YNB medium: 134 g yeast nitrogen base with ammonium sulfate without amino acids in 1 L of water, filter sterilize, *stable at 4°C for up to 1 year.*
14. 1.0 M sorbitol: *stable at 4°C for up to 1 year.*
15. Zeocin: *store at −20°C in the dark.*

3. Methods

3.1. Condensed Protocol for Making Electro-Competent *Pichia pastoris*

1. Aseptically pick a single *P. pastoris* colony from a freshly streaked YPD agar plate (see Note 1) into 5 mL of YPD medium. Culture overnight at 30°C with 250 rpm agitation.
2. Use the overnight culture to inoculate 50 mL of YPD medium in a 500-mL baffled flask to a final OD_{600} of 0.15–0.20.

3. Culture the yeast to an OD_{600} of 0.8–1.0 at 30°C with 250-rpm agitation.

4. Centrifuge the culture at $500 \times g$ for 5 min at room temperature and discard the medium.

5. Resuspend the cells in 9 mL of BEDS solution supplemented with 1 mL of 1 M DTT. This solution must be cooled on ice for a period of 30 min prior to use.

6. Incubate the cells at 30°C in a shaking incubator at 100 rpm for 5 min.

7. Centrifuge the sample at $500 \times g$ for 5 min at room temperature, discard the supernatant and resuspend the cells in 1 mL (0.02 volumes of the original 500 mL culture) of BEDS solution without DTT. Prepare 40 µL aliquots of the cells from this resuspension.

8. These cells are now electro-competent and ready for transformation. If the cells are not immediately required, the aliquots may be stored at −80°C for up to 6 months (do not snap freeze the cells in liquid nitrogen, directly store at −80°C).

3.2. Transformation of Electro-Competent Pichia pastoris *Cells*

1. Thaw a 40-µL aliquot of competent cells on ice (if previously frozen) per required transformation and transfer the cells into a 2-mm electroporation cuvette previously placed on ice for 5 min. Add 5 µg of linearized vector (with *Bst*XI, *Pme*I or *Sac*I) and incubate for 2 min on ice.

2. Electroporate the sample at 1,800 V with a 15-mS pulse length.

3. Resuspend the cells in 0.5 mL of 1.0 M sorbitol and 0.5 mL of YPD immediately after electroporation and incubate at 30°C in a shaking incubator at 100 rpm for 1 h (see Note 2). Plate the recovered cells on YPDS agar containing increasing concentrations of zeocin (100, 250, 500 or 1,000 µg/mL) for the selection of multicopy integrants.

4. Incubate the plates at 30°C for up to 1 week or until colonies are distinguishable.

5. Aseptically pick individual colonies and re-plate on YPDS agar with the same zeocin concentration as the initial selection plate and repeat the incubation in step 4 to generate a "Master" plate. This will then be used to screen for gene expression.

3.3. Making Competent Pichia pastoris *for PEG 1000 Transformation*

1. Aseptically pick a single *P. pastoris* colony from a freshly streaked YPD agar plate in to 5 mL of YPD medium. Culture overnight at 30°C with 250 rpm agitation.

2. Use the overnight culture to inoculate 100 mL of YPD medium in a 500-mL baffled shake flask to a final OD_{600} of 0.1 and culture at 30°C with 250 rpm agitation until an OD_{600} of 0.7 is obtained.

3. Centrifuge the culture at $5,000 \times g$ for 3 min at room temperature and resuspend the cells in 50 mL of BEDS solution without DMSO.

4. Centrifuge the cells at $5,000 \times g$ for 3 min, discard the supernatant, and resuspend the cells in 4 mL of BEDS solution without DMSO.

5. From this resuspension, prepare 200-µL aliquots in sterile 2-mL microfuge tubes, add 11 µL DMSO to each aliquot, immediately snap freeze in liquid nitrogen and store at −80°C.

3.4. Transformation of Pichia pastoris with PEG 1000

1. Add 1–50 µg of vector to the frozen aliquot of cells (see Note 3) *Do not thaw the cells.* Add 40 µg of sonicated salmon sperm DNA to the sample and incubate the tube in a 37°C water bath for 2 min.

2. Mix the sample by flicking the tube and incubate the sample as stated in step 1 for a further 3 min.

3. Remove the sample and add 1.5 mL of PEG 1000 solution, mix thoroughly by pipette, and incubate in a water bath at 30°C for 1 h.

4. Centrifuge the sample at $2,000 \times g$ for 10 min at room temperature.

5. Remove the supernatant and resuspend the pellet in 1.5 mL of Bicine Solution.

6. Centrifuge the sample as stated in step 4 and resuspend the cells in 200 µl of Bicine Solution.

7. Plate the 200-µL sample on to selective agar and incubate the plate at 30°C for up to 1 week.

3.5. Transformation of Saccharomyces cerevisiae Using Lithium Acetate

1. Aseptically pick a single *S. cerevisiae* colony from a freshly streaked YPD agar plate in to 10 mL of YPD medium and culture overnight at 30°C with 230-rpm agitation.

2. Centrifuge 0.5 mL of the culture in a 1.5-mL microcentrifuge tube at $5,000 \times g$ for 2 min at room temperature and discard the supernatant.

3. Add 1–5 µg of expression vector and 10 µL of sonicated salmon sperm DNA to the *S. cerevisiae* cells and resuspend by flicking the tube. Add 0.5 mL of Transformation Solution and vortex for 15–20 s to mix.

4. Incubate the tube without agitation at room temperature overnight (approximately 14 h).

5. Mix the sedimented cells with a pipette tip whilst withdrawing the bottom 80–100 µL from the tube and plate on selective agar. Incubate the plate at 30°C for 1 week or until individual colonies are distinguishable.

3.6. Transformation of Saccharomyces cerevisiae by Electroporation

1. Inoculate 10 mL of YPD with freshly streaked *S. cerevisiae* and culture overnight at 30°C, with an agitation speed of 250 rpm.

2. Inoculate 50 mL of YPD medium with the overnight culture to a final OD_{600} 0.1 and culture as above until an OD_{600} 1 is obtained.

3. Centrifuge the culture at $5,000 \times g$ for 3 min at room temperature.

4. Discard the medium and resuspend the cells in 25 mL of ice-cold MilliQ water.

5. Repeat steps 3 and 4.

6. Centrifuge the sample at $5,000 \times g$ for 3 min at room temperature, discard the supernatant, and resuspend the cells in 50 μL of 1.0 M sorbitol (see Note 4).

7. Remove a 40-μL aliquot of cells and place it in a 2-mm electroporation cuvette that has been left on ice for a period of 10 min prior to use. Add 5 μg of desalted DNA and 40 μg of sonicated salmon sperm DNA. Mix once by pipetting.

8. Incubate the sample on ice for 10 min.

9. Electroporate the sample at 1,800 V for 15 ms.

10. Immediately add 0.5 mL of 1.0 M sorbitol and 0.5 mL of YPD to the sample and incubate at 30°C in a shaking incubator at 100 rpm for 1 h.

11. Plate the recovered cells on selective media containing 1.0 M sorbitol and incubate at 30°C for 1 week or until colonies are distinguishable.

4. Notes

1. This plate should be no older than 5 days as this can lead to a reduction in the competence of the cells.

2. An increased number of transformants can be achieved by incubating the resuspended cells at 30°C with agitation for longer periods of time (1–3 h). However, this increase is due in part to replication of transformants and not necessarily due to increased transformation efficiency.

3. The volume of the vector added should not exceed 20 μL in total as this will lead to decreased transformation efficiency. A concentrated DNA sample will yield more transformants.

4. The efficacy of these cells decreases with the length of storage, and it is recommended that they are used immediately for optimum efficiency. If immediate use is not possible then the authors recommend that the cells are stored at 4°C and used within 3 days of preparation.

References

1. Baron M, Reynes JP, Stassi D, Tiraby G (1992) A selectable bifunctional beta-galactosidase:: phleomycin-resistance fusion protein as a potential marker for eukaryotic cells. Gene 114:239–243

2. Drocourt D, Calmels T, Reynes JP, Baron M, Tiraby G (1990) Cassettes of the *Streptoalloteichus hindustanus ble* gene for transformation of lower and higher eukaryotes to phleomycin resistance. Nucleic Acids Res 18:4009

3. Mulsant P, Gatignol A, Dalens M, Tiraby G (1988) Phleomycin resistance as a dominant selectable marker in CHO cells. Somat Cell Mol Genet 14:243–252

4. Perez P, Tiraby G, Kallerhoff J, Perret J (1989) Phleomycin resistance as a dominant selectable marker for plant-cell transformation. Plant Mol Biol 13:365–373

5. Lin-Cereghino J, Wong WW, Xiong S, Giang W, Luong LT, Vu J, Johnson SD, Lin-Cereghino GP (2005) Condensed protocol for competent cell preparation and transformation of the methylotrophic yeast *Pichia pastoris*. Biotechniques 38:44, 46, 48

6. Weiss HM, Haase W, Michel H, Reilander H (1998) Comparative biochemical and pharmacological characterization of the mouse 5HT(5A) 5-hydroxytryptamine receptor and the human beta(2)-adrenergic receptor produced in the methylotrophic yeast *Pichia pastoris*. Biochem J 330:1137–1147

Chapter 7

Screening for High-Yielding *Pichia pastoris* Clones: The Production of G Protein-Coupled Receptors as a Case Study

Shweta Singh, Adrien Gras, Cedric Fiez-Vandal, Magdalena Martinez, Renaud Wagner, and Bernadette Byrne

Abstract

Pichia pastoris is an established host for the production of a wide range of recombinant proteins including membrane proteins. The system has a particularly good track record for the production of G protein-coupled receptors (GPCRs). Generation and screening of expression clones with this system use standard molecular biology techniques. Multiple clones can be generated and screened in a matter of a few weeks making this similar to *Escherichia coli* in terms of speed. In addition, basic buffer components and the lack of expensive equipment make small-scale expression screening in *P. pastoris* very cost-effective. Here we describe the procedures used for small-scale GPCR production screening.

Key words: Small-scale expression, Western blot analysis, Functional analysis, Optimization trials

1. Introduction

Among the strains available for protein production in *P. pastoris*, the most popular and probably the best are the Mut⁺ strains, which grow on methanol as the sole carbon source. One such Mut⁺ strain, which has been used extensively for the production of GPCRs is the SMD1163 strain (*his4 pep4 prb1*) (1). The histidinol dehydrogenase gene (*HIS4*) has been mutated to allow the selection of successful transformants based on histidine auxotrophy. The SMD1163 strain is also a protease-deficient strain lacking the active genes coding for proteinase A (*PEP4*) and proteinase B (*PRB1*).

Roslyn M. Bill (ed.), *Recombinant Protein Production in Yeast: Methods and Protocols*, Methods in Molecular Biology, vol. 866, DOI 10.1007/978-1-61779-770-5_7, © Springer Science+Business Media, LLC 2012

Previous work has shown that use of this strain has resulted in higher yields of recombinant GPCRs, most probably as a result of reduced proteolytic degradation (2, 3). As mentioned in Chapter 3, generation of the *P. pastoris* expression plasmids involves straight-forward cloning of the appropriate genes into suitable restriction sites, the resulting constructs being used for yeast transformation. Since none of the available plasmids is able to autonomously replicate in *P. pastoris*, all of them, including pPIC9K, have to be integrated in the yeast genome through recombination events. The transformed clones containing the gene of interest are selected on the basis of a histidine prototrophy phenotype. Geneticin-resistant representative colonies are then placed in methanol-induced expression conditions to screen for the best-expressing clones as usually assessed by immunodetection experiments (3, 4). Once high-level expressing clones have been identified, these can be submitted to small-scale trials to allow assessment of the function of the recombinant receptor and to allow identification of optimized production conditions.

Yields of functional GPCRs in *P. pastoris* can be substantially improved by adjusting several experimental parameters such as the temperature and time of induction, the culture density at induction, the formulation of the growth medium and its supplementation with additives. Reducing the culture temperature during induction has been shown to have a positive effect on the functional yield of several GPCRs (4) and we routinely perform induction of receptor expression at 22°C (5, 6). The addition of receptor-specific ligand has also been shown to significantly increase GPCR yields, although this is receptor dependent (4, 7). For example, addition of ligand slightly increased production yields of functional $A_{2A}R$ (4). This may be because such ligands have been suggested to act as molecular chaperones, improving the efficiency of receptor folding (7, 8). Despite this, our typical protein preparation protocol for $A_{2A}R$ does not involve the addition of ligand (ZM241385) until the solubilization step, as reported by other researchers (9). This is because $A_{2A}R$ is a particularly stable GPCR; however, for other GPCRs, the addition of ligand during expression may be essential for production of sufficient quantities of stable receptor for downstream applications. The presence of the ligand can, however, complicate functional studies and, in the case of the $A_{2A}R$, we isolate the receptor in the absence of ligand for functional analysis.

Positive effects have also been observed following addition of dimethyl sulphoxide (DMSO) (3, 4, 10–12) to the expression culture. DMSO has been shown to facilitate phospholipid biosynthesis and membrane proliferation in yeast (13) and these features may result in a more suitable environment for the insertion of large amounts of receptor. One further additive shown to positively affect functional receptor expression is histidine. It is not clear precisely what the effects of histidine are although it has been suggested that it can act as an antioxidant (4). Further work is required in order to understand the precise role of histidine in receptor expression.

Overall, optimization of receptor expression can be highly successful, but is also often receptor dependent. Consequently, the systematic screening of expression conditions in a small-scale culture format is strongly recommended when initiating receptor expression trials using *P. pastoris*. Once optimal expression conditions have been identified, these can then be used to investigate alternative expression constructs. Here we provide protocols for both screening expression clones and some hints and tips on optimization of protein production (see Notes). The emphasis in this chapter is on procedures which worked very well for the production of the human adenosine A_{2A} receptor.

2. Materials

2.1. Generating Expression Colonies

1. Geneticin should be made as a stock solution at 100 mg/mL and filter-sterilized through a 0.2-μm syringe filter.

2. YPD plates: 1% (w/v) yeast extract, 2% (w/v) peptone, 2% (w/v) dextrose, 2% (w/v) agar supplemented with either 0.1 or 0.25 mg/mL geneticin. Autoclave 450 mL water containing 5 g yeast extract, 10 g peptone, and 10 g agar. Add 50 mL filter-sterilized 20% dextrose to the medium once it has cooled to 60°C. Follow the same procedure for YPG medium excluding agar.

3. YNB plates: 1.34% YNB, 2% dextrose, 1.5% agar, 0.00004% biotin. Autoclave 800 mL water containing 15 g agar. Add 100 mL filter-sterilized 13.4% YNB, 100 mL 20% dextrose, and 2 mL filter sterilized 0.02% biotin.

2.2. Small-Scale Expression

1. Buffered glycerol complex medium (BMGY): 100 mM potassium phosphate pH 6.0, 1% (w/v) yeast extract, 2% (w/v) peptone, 1.34% (w/v) yeast nitrogen base without amino acids, 0.00004% (w/v) biotin, 1% (w/v) glycerol. Prepare 700 mL of autoclaved 1% yeast extract, 2% peptone (YP media). Add 100 mL autoclaved 1 M potassium phosphate pH 6.0, 100 mL filter-sterilized 13.4% YNB, 2 mL filter-sterilized 0.02% biotin, and 100 mL autoclaved 10% glycerol to 700 mL of YP media.

2. Buffered methanol complex medium (BMMY): 100 mM potassium phosphate pH 8.0, 1% (w/v) yeast extract, 2% (w/v) peptone, 1.34% (w/v) yeast nitrogen base without amino acids, 0.00004% (w/v) biotin, 2.5% (v/v) dimethyl sulphoxide, 0.04% (w/v) histidine, and 0.3% (v/v) methanol.

3. YP medium: 1% yeast extract, 2% peptone in 800 mL water. Autoclave and add 100 mL autoclaved 1 M potassium phosphate pH 8.0, 100 mL filter-sterilized 13.4% YNB, 2 mL filter-sterilized 0.02% biotin, 2.5 mL of 100% methanol, and 1 mL filter-sterilized 4% histidine.

2.3. Small-Scale Membrane Preparation

1. Breaking buffer: 50 mM Na_2HPO_4 pH 7.5, 100 mM NaCl, 5% glycerol, 2 mM EDTA supplemented with protease-inhibitor tablets (Roche; 1 tablet/100 ml buffer).

2. Membrane buffer: 50 mM HEPES-NaOH, 100 mM NaCl, 10% (w/v) glycerol.

3. Methods

3.1. Preparation of Competent SMD1163 Cells

1. Inoculate 50 mL YPG medium with SMD1163 cells and incubate overnight at 30°C, 250 rpm (see Note 1).

2. Dilute the overnight culture in 400 mL YPG to $OD_{600} = 0.1$. Incubate the culture at 30°C, 250 rpm to an $OD_{600} = 1$.

3. Centrifuge the cells at $2,000 \times g$ for 5 min at 4°C.

4. Resuspend the cell pellet in 100 mL YPG, 20 mL HEPES 1 M pH 8, 2.5 mL 1 M DDT.

5. Incubate the cell suspension for 15 min at 30°C. Place on ice and make the volume up to 500 mL with ice-cold sterile water.

6. Centrifuge the cells at $2,000 \times g$ for 5 min at 4°C. Resuspend the pellet in 250 mL cold and sterile water.

7. Centrifuge the cells at $2,000 \times g$ for 5 min at 4°C. Resuspend the pellet in 20 mL of cold 1 M sorbitol.

8. Centrifuge the cells at $2,000 \times g$ for 5 min at 4°C. Resuspend the pellet in 500 μL of cold 1 M sorbitol.

9. Linearize 5–7 μg expression vector using 25 units of PmeI for 2 h at 37°C. Heat inactivate the enzyme by incubation at 60°C for 15 min.

3.2. Yeast Electrotransformation

1. Place an 800-μL electroporation cuvette (BioRad) on ice for at least 10 min.

2. Introduce 40 μL competent cells and 7.5 μL linearized DNA to the electroporation cuvette. Mix gently and incubate on ice for 5 min before proceeding with electroporation.

3. Set the electroporator to 1,500 V, 600 Ω, 25 μFD.

4. Place the cuvette in the electroporator and apply the electric pulse. Immediately resuspend the cells in 1 mL 1 M sorbitol.

5. Allow the cells to recover for about 1 h followed by centrifugation at $2,000 \times g$ for 10 min.

6. Discard the supernatant and resuspend the pellet in 500 μl 1 M sorbitol.

7. Plate 250 μL cell suspension on YNB agar plates and incubate for 2–3 days at 30°C.

3.3. Screening for Geneticin-Resistant Clones

The number of copies of the target gene which integrates into the *P. pastoris* genome can vary. It is possible to select clones using plates with increasing concentrations of geneticin (see also Chapter 6), since integration of multiple copies of the target gene also means integration of multiple copies of the geneticin-resistance gene. This can also be used to identify any false positives from the initial plating-out step.

1. Add 1 mL YPG medium onto the plates containing the His⁺ recombinant clones. Harvest by scraping the colonies from the agar plate using a T- or L-shaped spreader.

2. Prepare a serial dilution in YPG from 10 to 1,000-fold and measure OD_{600} for each sample.

3. Spread the equivalent of 10^5 cells/plate on YPG supplemented with 0.1 or 0.25 mg/mL geneticin. $OD_{600} = 1$ corresponds to 5×10^7 cells/mL.

4. Incubate for about 2–3 days at 30°C.

5. The colonies that appear on the plates are positive recombinant clones containing the gene. The colonies on plates containing 0.25 mg/mL geneticin will have a higher copy number of gene insertion than the ones on 0.1 mg/mL geneticin (see Note 2).

6. Select colonies for small-scale expression trials.

3.4. Expression of the Target Gene

1. Inoculate 5 mL BMGY medium with cells from a single high-copy-number colony.

2. Grow the cells overnight at 30°C in an incubator shaker to $OD_{600} = 12–15$. Use a 50-mL filter-capped Falcon tube which allows greater oxygenation of the yeast cell culture (see Note 3).

3. Centrifuge the cells at $3,000 \times g$ for 5 min.

4. Resuspend the cells in BMMY to $OD_{600} = 5$. Since recombinant protein expression is under control of the methanol inducible *AOX1* promoter, the methanol in the BMMY medium acts as the inducer.

5. Incubate the culture at 22°C for 18 h.

6. Harvest the cells by centrifugation at $3,000 \times g$ for 5 min. At this point the cells can either by snap-frozen in liquid N_2 and stored at −80°C or used immediately.

3.5. Preparation of Samples for Further Analysis: Small-Scale Membrane Preparation

1. Resuspend cells from a 10-mL culture in 500 μL ice-cold cell-breaking buffer.

2. Transfer the cell suspension together with 500-microlitreL glass beads to a 2-mL safe lock eppendorf tube.

3. Break the cells using a tissue lyser (Qiagen) set to 30 MHz for 15 min. In our laboratory the tissue lyser is kept in the cold room to minimize heating of the protein samples during breakage. If a mechanical tissue lyser is not available, then it is possible to break the cells using a vortex mixer in the cold room.

4. Centrifuge the sample at $1,500 \times g$ for 1 min to separate the beads. Transfer the solution into a clean 1.5-mL eppendorf tube.

5. Centrifuge the sample at $3,000 \times g$ for 10 min in order to pellet unbroken cells. Transfer the supernatant into a clean 1.5-mL eppendorf tube.

6. Centrifuge the supernatant at $100,000 \times g$ for 30 min. For this step we use an Optima Max benchtop ultracentrifuge (Beckman) with a TLA 55 rotor.

7. Resuspend the pellets in ice-cold membrane buffer. The samples can then be either flash-frozen in liquid N_2 or used immediately for expression or functional analysis (see Note 4).

3.6. Expression and Functional Analysis

The easiest way to assess expression is to perform Western blot analysis using affinity-tag-specific antibodies. The vector system we use has been modified to incorporate amino-terminal Flag and His tags, although a number of other detection tags are available (see Chapter 3). High-quality antibodies against Flag and His are readily available. It is possible to harvest samples at different time points or after induction at different temperatures, for example, and compare the Western blot signals. However, this is only ever a qualitative analysis and does not differentiate between functional and nonfunctional protein. Radioligand binding should be used to assess the functional yields of GPCRs (see Note 5). Single point binding assays can give a rapid assessment of the functional yields from a number of different conditions (Fig. 1).

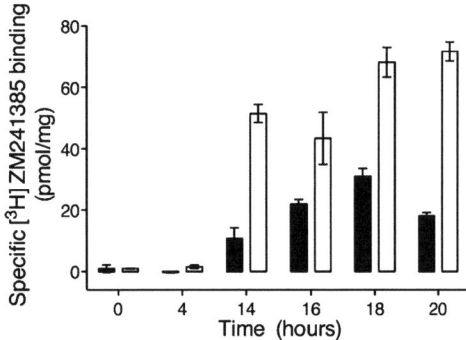

Fig. 1. Functional yields of two A$_{2A}$R constructs over time. Radioligand binding analysis of human wild-type adenosine A$_{2A}$ (*filled bars*) and a carboxy-terminal truncated A$_{2A}$R (*open bars*) receptors produced in small-flask (100 mL) cultures. Radioligand binding assay using [^{3}H] ZM241385 was performed on membranes prepared from cells harvested at different time points. Induction of the cultures was initiated by addition of methanol at time 0 h. The cultures were induced with 0.3% methanol for 20 h. Data shown are representative of at least duplicate experiments for each condition. The data suggest that the optimal time to harvest cells is after 18-h induction.

4. Notes

1. The pPIC9K expression vector in combination with the SMD1163 cells has been well characterized for GPCR production although other vectors and cell strains are available. The mammalian glucose transporters, GLUT1 and GLUT4 (14), and a human aquaporin 1 (15) have been successfully produced to high levels using the pPICKZ vector (Invitrogen, Carlsbad) in X-33 cells. In addition it has proved possible to produce the human peripherin/RDS protein and a range of human ABC transporters using the pPICKZ vector in combination with the KM71H strain (16). The techniques used in each case for generation and selection of the expression clones are very similar to that outlined here.

2. While screening clones for copy number can be useful, it should be noted that clones containing the highest copy number do not necessarily give the highest production yields (3, 4). It is important to further screen using Western blot and functional analysis to identify the best expressing clones.

3. Small-scale cultures (10 mL) can be performed in 50-mL conical tubes; however, aeration may not be optimal and it may be preferable to scale up cultures to 100-mL baffled flask trials to maximize expression level and increase reproducibility of the culture conditions.

4. It is possible to scale up using large flask cultures; however, it is very important to use a relatively small volume/flask ratio (200 mL/2 L) in order to ensure adequate oxygenation of the culture.

5. A combination of Western blot and functional analysis allows an estimate of the yield of functional versus nonfunctional receptor protein.

Acknowledgments

This research was funded by the MepNet consortium, BBSRC, and GlaxoSmithKline.

References

1. Gleeson MA, White CE, Meininger DP, Komives EA (1998) Generation of protease-deficient strains and their use in heterologous protein expression. Methods Mol Biol 103:81–94

2. Weiss HM, Haase W, Michel H, Reilander H (1995) Expression of functional mouse 5-HT5A serotonin receptor in the methylotrophic yeast *Pichia pastoris*: pharmacological characterization and localization. FEBS Lett 377:451–456

3. Weiss HM, Haase W, Michel H, Reilander H (1998) Comparative biochemical and pharmacological characterization of the mouse 5HT5A 5-hydroxytryptamine receptor and the human beta2-adrenergic receptor produced in the methylotrophic yeast *Pichia pastoris*. Biochem J 330:1137–1147

4. André N, Cherouati N, Prual C, Steffan T, Zeder-Lutz G, Magnin T, Pattus F, Michel H, Wagner R, Reinhart C (2006) Enhancing functional production of G protein-coupled receptors in *Pichia pastoris* to levels required for structural studies via a single expression screen. Protein Sci 15:1115–1126

5. Singh S, Gras A, Fiez-Vandal C, Ruprecht J, Rana R, Martinez M, Strange PG, Wagner R, Byrne B (2008) Large-scale functional expression of WT and truncated human adenosine A2A receptor in *Pichia pastoris* bioreactor cultures. Microb Cell Fact 7:28–38

6. Singh S, Hedley D, Kara E, Gras A, Iwata S, Ruprecht J, Strange PG, Byrne B (2010) A purified C-terminally truncated human Adenosine A_{2A} receptor construct is functionally stable and degradation resistant. Protein Expr Purif 74:80–87

7. Grunewald S, Haase W, Molsberger E, Michel H, Reilander H (2004) Production of the human D2S receptor in the methylotrophic yeast, *P. pastoris*. Receptors Channels 10:37–50

8. King K, Dohlman HG, Thorner J, Caron MG, Lefkowitz RJ (1990) Control of yeast mating signal transduction by a mammalian β2-adrenergic receptor and Gs α subunit. Science 250:121–123 [Erratum: Science 1991 251, 144]

9. Jaakola VP, Griffith MT, Hanson MA, Cherezov V, Chien EYT, Lane JR, IJzerman AP, Stevens RC (2008) The 2.6 angstrom crystal structure of a human A2A adenosine receptor bound to an antagonist. Science 322:1211–1217

10. Sarramegna V, Talmont F, Demange P, Milon A (2003) Heterologous expression of G-protein-coupled receptors: comparison of expression systems from the standpoint of large-scale production and purification. Cell Mol Life Sci 60:1529–1546

11. Fraser NJ (2006) Expression and functional purification of a glycosylation deficient version of the human adenosine 2a receptor for structural studies. Protein Expr Purif 49:129–137

12. Shukla AK, Haase W, Reinhart C, Michel H (2007) Heterologous expression and characterization of the recombinant bradykinin B2 receptor using the methylotrophic yeast *Pichia pastoris*. Protein Expr Purif 55:1–8

13. Murata Y, Watanabe T, Sato M, Momose Y, Nakahara T, Oka S, Iwahashi H (2003) Dimethyl sulfoxide exposure facilitates phospholipid biosynthesis and cellular membrane

proliferation in yeast cells. J Biol Chem 278: 33185–33193

14. Alisio A, Mueckler M (2010) Purification and characterization of mammalian glucose transporters expressed in *Pichia pastoris*. Protein Expr Purif 70:81–87

15. Nyblom M, Oberg F, Lindkvist-Petersson K, Hallgren K, Findlay H, Wikström J, Karlsson A, Hansson O, Booth PJ, Bill RM, Neutze R,

Hedfalk K (2007) Exceptional overproduction of a functional human membrane protein. Protein Expr Purif 56:110–120

16. Vos WL, Vaughan S, Lall PY, McCaffrey JG, Wysocka-Kapcinska M, Findlay JBC (2010) Expression and structural characterization of peripherin/RDS, a membrane protein implicated in photoreceptor outer segment morphology. Eur Biophys J 39:679–688

Chapter 8

Screening for High-Yielding *Saccharomyces cerevisiae* Clones: Using a Green Fluorescent Protein Fusion Strategy in the Production of Membrane Proteins

David Drew and Hyun Kim

Abstract

The overproduction of eukaryotic membrane proteins in milligram quantities is a major bottleneck for their further biochemical and structural investigation. Production trials exploring a range of input factors can be rationalized to improve the likelihood of success. Here we discuss some of these factors in combination with the use of a GFP-based *Saccharomyces cerevisiae* system that enables a quick turnaround time from clone construction to production trials. Since membrane-integrated levels do not necessarily correlate with the amount of functional recombinant protein, we also include the use of fluorescence-detection size exclusion chromatography (FSEC). Using FSEC, the quality of the recombinant material can also be rapidly evaluated as demonstrated for the functional production of the rat vesicular glutamate transporter (VGLUT2) and the human glucose transporter (GLUT1) (5).

Key words: Membrane protein, Overproduction, *S. cerevisiae*, Fluorescence-detection size exclusion chromatography

1. Introduction

Membrane protein overproduction is an empirically based approach where many parameters need to be tested and re-tested. Monitoring by fluorescence enables this process to be carried out quickly, efficiently, and reliably. For this purpose, we use GFP-based fusion technology. As a carboxy-terminal GFP tag will only fold and becomes fluorescent if the upstream membrane protein integrates into the membrane, the resultant fluorescence is a fast and accurate measure of membrane-integrated production (1). Fluorescence is easy to measure directly in liquid culture, standard SDS-gels and

Roslyn M. Bill (ed.), *Recombinant Protein Production in Yeast: Methods and Protocols*, Methods in Molecular Biology, vol. 866, DOI 10.1007/978-1-61779-770-5_8, © Springer Science+Business Media, LLC 2012

in detergent-solubilized membranes (2, 3). Detergent-solubilized membranes can also be further subjected to fluorescence size-exclusion chromatography (FSEC) to measure the "monodispersity" of the sample. This is an ideal way to evaluate the quality of the material produced. In short, although the amount of membrane-integrated production is no guarantee that the recombinant protein is functional, the GFP-tag speeds up the empirical process.

We have constructed a reliable protocol for screening the overproduction and purification of eukaryotic membrane proteins in *Saccharomyces cerevisiae* (4). This system was adapted from a GFP-based *Escherichia coli* pipeline (3) because yeast possess features absent in *E. coli* that are often essential for producing functional eukaryotic membrane proteins. With similar costs to *E. coli* and the possibility to clone into standard vectors by homologous recombination, *S. cerevisiae* is a convenient and efficient production host. For this reason we prefer to screen and optimize in this yeast rather than *Pichia pastoris*. In addition, because final cell densities are generally in the range of 6.0–8.5 at OD_{600}, rather than around 60–80 for *P. pastoris*, production of 1 mg/L recombinant protein in *S. cerevisiae* represents a larger fraction of total protein that would be the case in *P. pastoris*: a higher fraction of recombinant protein ensures better recovery of it in a purer form. From an analysis of ~150 eukaryotic membrane protein–GFP fusions, we found that around one quarter can be overproduced to >1 mg/L. Of the highly produced eukaryotic membrane proteins in *S. cerevisiae*, more than half of those tested were targeted to the correct organelle and were monodisperse in a mild detergent such as dodecyl-β-D-maltopyranoside (DDM).

Here, we describe in detail the practical steps that constitute our *S. cerevisiae* GFP-based pipeline. This comprises (1) cloning by homologous recombination, (2) whole-cell and in-gel fluorescence for estimating production yields and (3) FSEC for judging the quality of the recombinant material. In Chapter 18, we expand on these methods for large-scale production and purification.

2. Materials

1. Expression vectors (see Chapter 4).
2. Materials for yeast transformation (see Chapter 4).
3. PCR reagents (available from a wide range of suppliers).
4. *Sma*I restriction enzyme (Invitrogen).
5. Growth medium without uracil (for 1 L, 6.7 g yeast nitrogen base without amino acids (BD Difco, cat. No. 291920)), 2 gt yeast synthetic drop-out medium supplement without uracil

(Sigma, cat. No. Y1501), and either 2% glucose (for pre-culture) or 0.1% glucose (for expression culture)). For plates, add 20 g of bacto agar (Sigma, cat No. A5306). D-(+) glucose can be purchased from Sigma (cat. No. G7021).

6. 20% galactose (w/v) (Sigma).

7. YSB (yeast suspension buffer): 50 mM Tris–HCl (pH 7.6), 5 mM EDTA, 10% glycerol, 1× complete protease inhibitor cocktail tablets.

8. Nunc 96-well black optical-bottom plates (Nunc).

9. SpectraMax M2e microplate reader (Molecular Devices).

10. Acid-washed glass beads, 500 μm (Sigma).

11. TissueLyser mixer (Qiagen).

12. Benchtop ultracentrifuge, Beckman Coulter Optima MAX series with TLA-55 and TLA-120.1 rotors (Beckman).

13. 1.5-mL polyallomer microcentrifuge tubes (Beckman).

14. SB (sample buffer) for in-gel fluorescence: 50 mM Tris–HCl (pH 7.6), 5% glycerol, 5 mM EDTA (pH 8.0), 4% SDS, 50 mM DTT, 0.02% bromophenol blue.

15. Tris-glycine SDS gels.

16. Fluorescent protein standard (Invitrogen).

17. Pre-stained protein standard (Invitrogen).

18. LAS-1000-3000 charge-coupled device (CCD) imaging system (Fujifilm).

19. Coomassie brilliant blue R-250 (Sigma).

20. CRB (cell resuspension buffer): 50 mM Tris–HCl (pH 7.6), 1 mM EDTA, 0.6 M sorbitol.

21. Constant Systems TS series cell disruptor (Constant Systems).

22. MRB (membrane resuspension buffer): 20 mM Tris–HCl (pH 7.6), 0.3 M sucrose, 0.1 mM $CaCl_2$.

23. Bicinchoninic acid (BCA) protein assay kit (Pierce).

24. PBS (phosphate buffer saline): For 1 L, 1.44 g $Na_2HPO_4 \cdot 2H_2O$ (8.1 mM phosphate), 0.25 g KH_2HPO_4 (1.9 mM phosphate), 8 g NaCl, 0.2 g KCl, adjust pH to 7.4 using 1 M NaOH or HCl.

25. Dodecylnonaoxyethylene ether (C12E9; Anatrace).

26. *N,N*-Dimethyldodecylamine *N*-oxide (LDAO; Anatrace).

27. Cholesteryl hemisuccinate; Tris salt (CHS; Sigma).

28. *n*-Dodecyl-β-D-maltopyranoside (DDM; Anatrace).

29. Superose 6 10/300 GL Tricorn gel filtration column (GE Healthcare).

30. Äkta FPLC system (GE Healthcare).

31. Frac-950 fraction collector with rack C (GE Healthcare).

32. 50 mL aerated capped tubes (Techno Plastic Products (TPP)).

3. Methods

3.1. Rationalizing the Construct Design and Cloning of Membrane Protein-Encoding Gene(s) into a GFP-His$_8$-Containing Vector

1. Analyze the membrane protein sequence for regions of disorder using the algorithm RONN (http://www.strubi.ox.ac.uk/RONN). Consider designing amino- or carboxy-terminal truncations based on this output. To guide construct design, compare the analysis with the known or predicted topology (see TMHMM (http://www.cbs.dtu.dk/services/TMHMM/)), and also to sequence alignment with close homologues by ClustalW (http://www.ebi.ac.uk/Tools/clustalw2/)). See Note 1.

2. As outlined in Chapter 4, create a 2μ vector that codes for yEGFP-His$_8$ and contains a site for protease cleavage, e.g. TEV protease (see Note 2).

3. Amplify the cDNA clone of interest with primers that contain approximately 35-bp complementary 5′ overhangs to the SmaI-linearized GFP-fusion vector (see Chapter 4).

4. Transform S. cerevisiae competent cells with 3 μL PCR product and 5 μL linearized vector (see Note 3).

3.2. Measuring Yields by Whole-Cell and In-Gel Fluorescence

1. Inoculate 10 mL growth medium without uracil plus 2% glucose with a single colony in an aerated 50-mL tube (see Note 4).

2. Incubate the culture overnight in an orbital shaker at 30°C, 280 rpm.

3. Spot 10 μL of the overnight culture onto a fresh plate without uracil, allow the spot to dry at room temperature and transfer the plate to a 30°C incubator for 1–2 days.

4. Dilute the overnight culture (from step 2) to OD_{600} 0.12 in two 50-mL aerated tubes, each containing 10 mL growth medium without uracil plus 0.1% glucose (see Note 5).

5. Incubate the cultures in an orbital shaker at 30°C, 280 rpm. At OD_{600} 0.6 (after approximately 7 h), induce production of the membrane protein–GFP fusion by adding 20% (w/v) galactose to achieve a final concentration of 2% (see Note 6).

6. 22 h post-induction, centrifuge the cells at $3,000 \times g$ for 5 min, remove the supernatant and resuspend the cell pellet in 200 μL YSB (see Note 7).

Table 1
Membrane protein yield estimates from whole cells

1. Harvest 10 mL yeast cells that have been cultured with and without galactose addition (to estimate background fluorescence), remove supernatant and resuspend in 200 μL YSB

2. Measure the GFP fluorescence
   For example, with no galactose (MP-GFP − GAL) = 3,000 relative fluorescence units (RFU). With galactose (MP-GFP + GAL) = 32,000 RFU

3. Correlate the whole-cell fluorescence with the amount of GFP produced by measuring the fluorescence of a defined concentration of yeast-enhanced green fluorescent protein (yEGFP) in a final volume of 200 μL
   For example, in our plate reader, fluorescence of pure yEGFP at a concentration of 0.03 mg/mL is 11,300 RFU

4. Determine the concentration of GFP in 200 μL cell culture by dividing by 40 (i.e., 8,000 μL (cell culture)/200 μL (resuspension volume)). Note that although the initial cell culture volume was 10 mL, there is an effective 2 mL loss through the transfer of only 200 μL of the resuspended cells (200 μL buffer + cell pellet = 250 μL) to the 96-well plate
   Using the above example: ((32,000 − 3,000)/11,300) × 0.03/40 = 0.001 9 mg/mL, which equates to an expression yield of 1.9 mg/L

5. As the typical recovery of GFP fluorescence from 1 L culture into membranes is 60%, multiply the above number by 0.6:
   1.9 mg/L × 0.6 = 1.1 mg/L

6. To calculate the amount of membrane protein, multiply the above number by the molecular weight of the membrane protein/GFP (28 kDa)
   For example: a 56-kDa membrane protein with 32,000 RFU;
   1.1 mg/L × (56/28) kDa = 2.2 mg/L

7. Transfer 200 μL of the cell suspension to a black Nunc 96-well optical-bottom plate (see Note 8).

8. Measure the GFP fluorescence emission at 512 nm following excitation at 488 nm in a microplate spectrofluorometer. For plate readers with a bottom-read option, choose this setting. Estimate membrane protein yield (in mg/L) from the yeast whole-cell fluorescence reading by applying the methodology detailed in Table 1.

9. Transfer the cell suspension from the 96-well plate into a 1.5-mL capped tube.

10. Add glass beads so that the final volume including the cell suspension is 500 μL. Add an additional 500 μL YSB.

11. Break the yeast cells with a mixer-mill disruptor set at 30 Hz for 7 min at 4°C. Alternatively, a vortexer can be used, but we recommend using a heavy-duty disruptor, as cell breakage is more efficient, reproducible, and easier to scale up.

12. Remove unbroken cells by centrifugation at $22,000 \times g$ in a desktop centrifuge for 5 s at 4°C. Transfer 500 μL supernatant into a new tube. Add 500 μL YSB to the mixture of unbroken cell pellet and glass beads. Repeat step 11 and transfer the supernatant to the 500 μL batch obtained from the first round of cell breakage.

13. To pellet crude membranes, centrifuge the 1 mL supernatant from step 12 at $20,000 \times g$ in a desktop centrifuge at 4°C for 1 h. Alternatively, the supernatant can be centrifuged using a desktop ultracentrifuge ($120,000 \times g$ for 1 h). However, we find the recovery from centrifugation in a desktop centrifuge is sufficient for this analysis and, as the final pellet is less compact, it is easier to resuspend (step 14).

14. Resuspend crude membranes in 50 μL YSB and transfer 15 μL into a tube containing 15 μL SB. Load 10 μL for SDS-PAGE. Include non-fluorescent and fluorescent protein standards, such as Benchmark Fluorescent and SeeBlue Plus Prestained standards (both from Invitrogen), respectively. For this step, we recommend our SB composition with standard SDS denaturing cast gels for in-gel fluorescence. We have also tested pre-cast Criterion (Bio-Rad) and Tris–Gly gels with equal success. We have found that the NuPAGE gels (Invitrogen) are not compatible with in-gel fluorescence (see Note 9).

15. Rinse the SDS gel with de-ionized H_2O and detect the fluorescent bands with a CCD camera system. Expose the gel to blue light (EPI source) set at 460 nm with a cut-off filter of 515 nm. Capture images and increase the exposure time until the fluorescent bands are clearly visible (see Note 10).

16. Analyze the gel and compare the size of the bands to the protein standards. If two closely spaced bands are present, this could indicate that the protein is glycosylated (see Fig. 1). In this case, analyze the sequence for N-linked glycosylation sites (http://www.cbs.dtu.dk/services/NetNGlyc/; see Note 11).

17. Stain the gel with Coomassie Brilliant Blue and transfer to destain (see Note 12).

3.3. Estimating the Quality of the Recombinant Protein by FSEC

1. Inoculate 10 mL growth medium without uracil with the spotted yeast culture from step 3 of Subheading 3.2 and incubate overnight.

2. The next day, add the overnight culture to a 500-mL shake flask containing 150 mL growth medium without uracil and

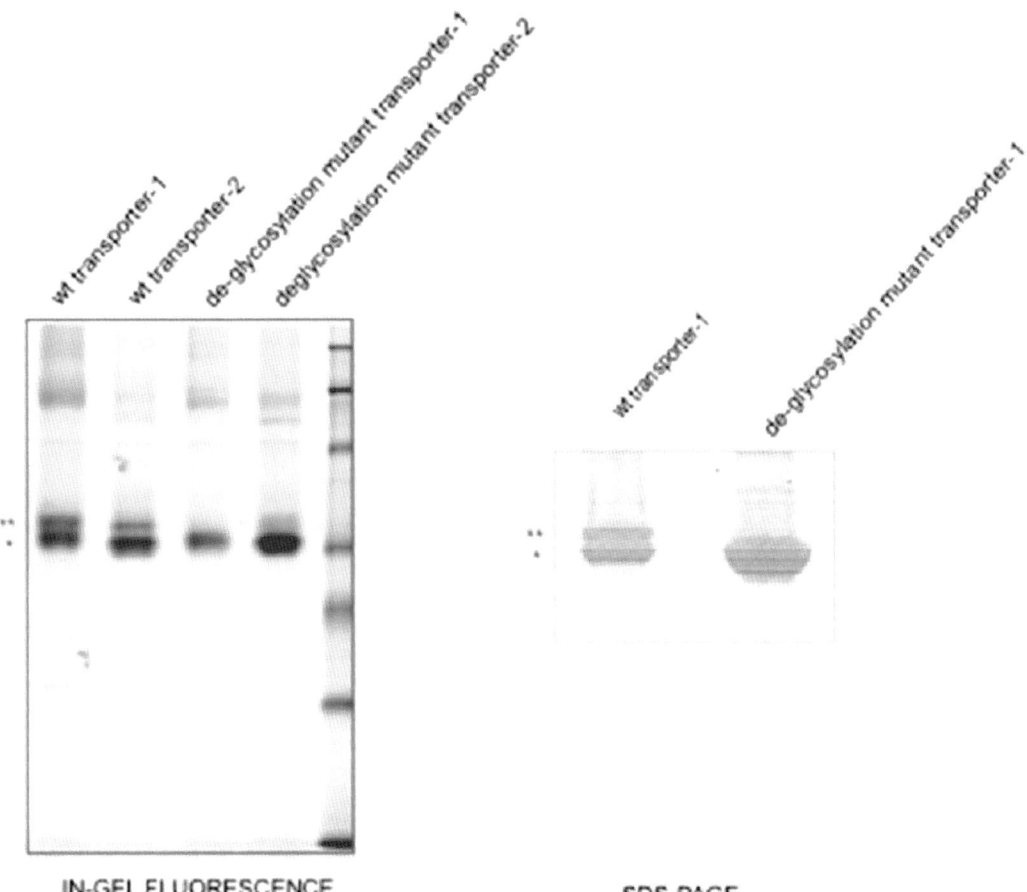

Fig. 1. Example of N-linked glycosylation of recombinant mammalian transporter-GFP fusion from *S. cerevisiae*. *Left panel*: SDS/PAGE and in-gel fluorescence detection in crude membranes of mammalian transporter-1 and -2 (lanes 1 and 2) and, after mutation of asparagine to alanine in the N-X-S/T motif, of the same transporters (lanes 3 and 4): *single and double asterisks* represent unglycosylated and glycosylated proteins, respectively. *Right panel*: Coomassie staining of the purified transporter before and after asparagine mutation of the mammalian transporter-1, as illustrated in the *left panel*.

2% glucose. Incubate the culture overnight in an orbital shaker at 280 rpm, 30°C.

3. Dilute the 150 mL overnight culture to OD_{600} 0.12 in 1 L growth medium without uracil containing 0.1% glucose in a 2.5-L baffled shake flask. Incubate the culture in an orbital shaker at 280 rpm, 30°C.

4. Harvest the cells after 22 h by centrifugation at $4,000 \times g$ at 4°C for 10 min. Decant the supernatant and resuspend the cell pellet in 25 mL CRB/L original cell culture.

5. Disrupt the cells with four passes in a heavy-duty cell disruptor at incremental pressures of 25, 30, 32 and 35 kpsi (approximately $1.7–2.4 \times 10^3$ atm) at 4–15°C. Remove 100 µL cells, transfer into a 96-well plate and measure GFP fluorescence as outlined in step 8 of Subheading 3.2.

6. Remove the unbroken cells and debris by centrifugation at 10,000 × *g* at 4 °C for 10 min and collect the supernatant, which contains the membrane fragments. Transfer 100 μL supernatant to a 96-well plate and measure GFP fluorescence as outlined in step 8 of Subheading 3.2. Calculate the yeast cell breakage efficiency by comparing the GFP fluorescence to that measured in step 5 (see Note 13).

7. To collect the membranes, centrifuge the cleared supernatant at 150,000 × *g* at 4 °C for 120 min. Discard the supernatant and resuspend the pellet to a final volume of 6 mL MRB/L original cell culture using a disposable 10-mL syringe with a 21-gauge needle. Transfer 100 μL membrane suspension to a 96-well plate and measure GFP fluorescence as outlined in step 8 of Subheading 3.2. Calculate the amount of recombinant membrane protein (see Table 1). Calculate the amount of total protein using the BCA protein assay kit following the manufacturer's instructions (see Note 14).

8. Adjust the volume of the membrane suspension to achieve a protein concentration of 3.5 mg/mL in PBS. Transfer 900-μL aliquots of this membrane suspension into 1.5-mL Beckman polyallomer microcentrifuge tubes.

9. Add 100 μL freshly prepared 10% (w/v) stock of $C_{12}E_9$, 12 M, 10 M, 9 M, or LDAO to the 1.5-mL tubes containing 900 μL membrane suspension (achieving a final concentration of 1% detergent and a final protein concentration of 3.2 mg/mL). Incubate the mixtures at 4 °C for 1 h with mild agitation. We recommend testing the addition of cholesteryl hemisuccinate (CHS) to the detergent mixture (at a final concentration of 0.2%), as this can be essential for the isolation of monodisperse mammalian membrane proteins.

10. Transfer 100 μL detergent-solubilized membrane protein solution into a 96-well plate and measure GFP fluorescence as outlined in step 8 of Subheading 3.2. To remove the non-solubilized material, centrifuge the remaining 900 μL in a benchtop ultracentrifuge at 100,000 × *g* at 4 °C for 45 min.

11. Transfer the clarified supernatant to a new 1.5-mL tube. Transfer 100 μL to a 96-well plate and repeat the GFP fluorescent measurement as outlined in step 8 of Subheading 3.2. Calculate the detergent solubilization efficiency by comparing the GFP fluorescence measurement with that in step 7 (see Note 15).

12. Inject 0.5 mL detergent-solubilized sample onto a Superose 6 10/300 column equilibrated in 20 mM Tris–HCl (pH 7.5), 0.15 M NaCl and 0.03% DDM. After elution of the first 6 mL, collect 0.2-mL fractions row by row into a 96-well plate (see Note 16).

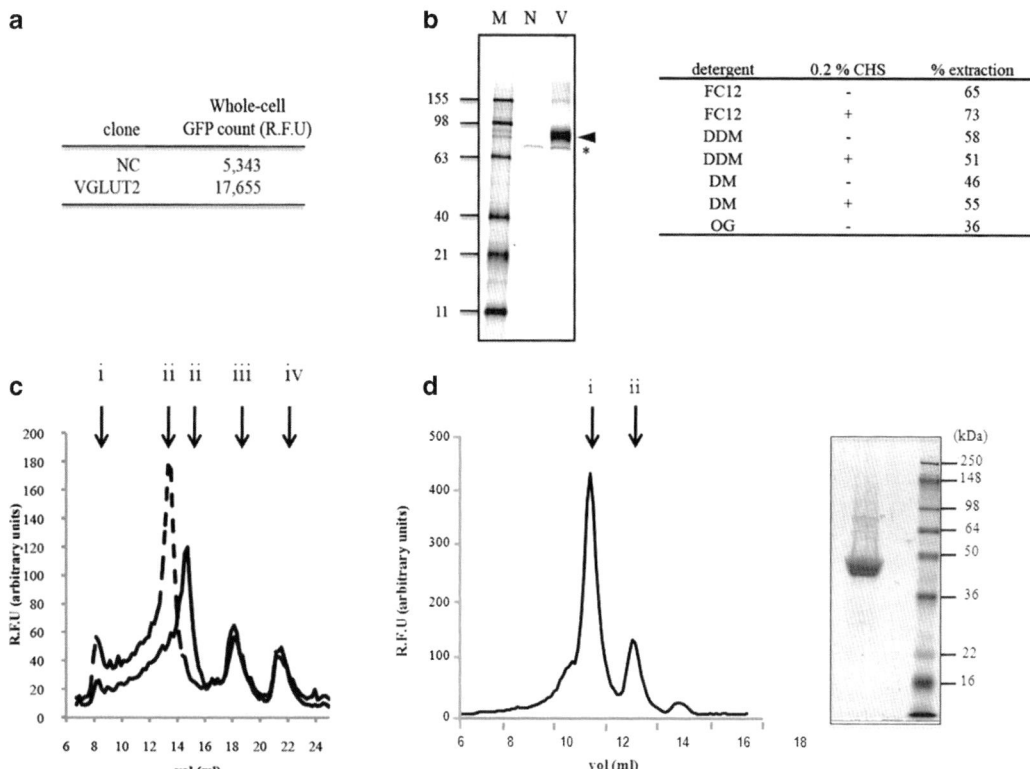

Fig. 2. Example of overproduction screening of rat VGLUT-2 and human GLUT-1. (a) Table illustrating the amount of whole-cell fluorescence from cells producing VGLUT-2 and a non-induced control (as outlined in Table 1, this information is used to calculate yields in mg/L). (b) *Left panel*: SDS-PAGE and in-gel fluorescence detection of crude membranes; lane M = Benchmark™ fluorescent molecular weight marker, lane N = non-induced cells, lane V = induced cells (the *asterisk* indicates endogenous fluorescent protein and the *arrow* indicates VGLUT2-GFP production) *Right panel*: Solubilization efficiency of membranes containing VGLUT-2 using Fos-choline 12 (FC12), dodecyl-b-D-maltopyranoside (DDM), decyl-β-D-maltopyranoside (DM) in the presence or absence of 0.2% cholesterol hemisuccinate (CHS). (c) FSEC trace of DM-solubilized VLGUT2 membranes in the absence (*solid line*) and presence of CHS (*dashed line*); i = aggregation peak, ii = VGLUT2-GFP fusion peak, iii = free GFP peak, iv = endogenous fluorescent protein. *Note*: as outlined in ref. 5, although solubilization was higher in other detergents, DM with CHS gives the best monodisperse profile and was used in the isolation of functional VGLUT-2. (d) *Left panel*: FSEC trace of DDM-solubilized human GLUT-1; i = GLUT1-GFP, ii = free GFP peak, *right panel*: SDS-PAGE analysis of purified human GLUT-1 in DDM as detected by Coomassie staining. *Note*: as outlined in ref. 5, recombinant human GLUT-1 transports D-glucose comparably with GLUT-1 purified from its native source.

13. Set the 96-well parameters of the plate reader to read wells row by row. To improve the signal-to-noise ratio, measure the GFP fluorescence emission at 512 nm by excitation at 470 nm (this wavelength is used instead of 488 nm, as it produces a lower background fluorescent signal). Plot the GFP fluorescence in each well against the fraction number (see Note 17).

14. Analyze the FSEC trace. As shown in Fig. 2, a monodisperse protein–detergent complex peak is symmetrical and elutes within a total volume of 2 mL. Larger volumes are not characteristic of stable and/or well-folded proteins in detergent (see Note 18).

4. Notes

1. Eukaryotic membrane proteins typically contain regions that are predicted by their amino acid sequence to be disordered. Most often these regions are in large loops or the amino- or carboxy-terminal tails. If a disordered region is protein-specific it is less likely to be conserved or contain a transmembrane (TM) segment. In many cases, minor amino- or carboxy-terminal truncations have been shown to improve stability (6, 7).

2. Although we recommend a *GAL1* promoter, in some cases, a constitutive promoter gives higher yields. As there are examples where an amino-terminal GFP fusion is more suitable for functional or structural work than a carboxy-terminal GFP fusion, an amino-terminal GFP fusion vector should also be considered. However, an amino-terminal GFP fusion may reduce production for N_{out}-membrane protein topologies as it may interrupt amino-terminal translocation, and as the upstream GFP can be translated more efficiently than the downstream membrane protein, the GFP fluorescence from whole cells may no longer be a reliable reporter of membrane-integrated production.

3. It is recommended that at least ten close homologues are initially tested to maximize the use of this approach, since the typical success rate for the production of mammalian membrane proteins in *S. cerevisiae* is 20%. We use a strain that has the gene encoding the vacuolar Pep4 protease deleted. The genotype should be compatible with the vector selection marker used.

4. Aerated capped tubes should be used, as they allow a more reliable estimate of yields at a large scale than non-aerated tubes.

5. 0.1% glucose, and not 2% glucose, is used in the production medium because high levels of glucose repress the *GAL1* promoter, while the former helps to maintain cell growth.

6. In order to avoid diluting the culture medium, prepare the 20% galactose stock in medium lacking uracil. Induction before OD_{600} 0.6 (typically) causes a reduced biomass that lowers protein yields. Although whole-cell GFP fluorescence can be higher if the cells are induced at a higher OD_{600} than 0.6, we find that there is proportionately greater degradation, so there is no linear gain in the amount of membrane-integrated material produced.

7. It is important to culture for 22 h after galactose addition. Although the OD_{600} is constant after 12 h, membrane protein production is maximal at 12–20 h post-induction (8). Because different final volumes can affect the level of whole-cell fluores-

cence measured, it is important to remove all the supernatant. We recommend removal of supernatant by vacuum suction, or if it is removed by hand, by patting the tubes dry using absorbent paper.

8. Because yeast cells settle to the bottom of the plate, proceed to the next step within 5 min of transfer to obtain accurate measurements.

9. Do not boil samples for SDS-PAGE, as this denatures GFP and often causes membrane protein samples to aggregate.

10. Blue light is recommended over UV light, as it is closer to the excitation wavelength of GFP. In addition, we do not recommend detecting GFP–fusion production by Western blotting, as the transfer of membrane protein–GFP fusions is often inconsistent among membrane protein samples.

11. Asn-X-Ser/Thr N-linked glycosylation acceptor sequences must be on the luminal side of the ER to be glycosylated by oligosaccharyltransferase and a distance of about 12–13 amino acids away from the end of the TM segment.

12. The intensity of Coomassie Brilliant Blue staining is a poor indication of the yield, as some membrane proteins bind the dye better than others.

13. Breakage efficiency should be greater than 80%. If lower than this, consider diluting the sample before breakage to improve efficiency.

14. If the membranes isolated from 1 L of cells are resuspended in 6 mL MRB, the GFP fluorescent counts typically match the original whole-cell fluorescent counts (8). This corresponds to approximately 60% of the amount of GFP measured in whole cells being incorporated into membranes. Note, membrane suspensions can be rapidly frozen in liquid nitrogen and stored at –80°C for up to 6 months. Although this is the routine in our laboratory, some membrane protein crystallographers avoid freezing and storing membranes and continue with purification immediately.

15. The detergent solubilized supernatant can be rapidly frozen in liquid nitrogen and stored at –80°C. Note, however, that some membrane protein crystallographers do not freeze and analyze by FSEC directly.

16. The use of a low percentage of DDM (0.03%) in the buffer used for separation of the detergent-solubilized membranes by SEC does not rescue membrane protein that has aggregated in the original detergent.

17. Ideally, as originally outlined by Kawate and Gouaux, fluorescence can be measured with higher sensitivity using an in-line detector, connected directly to the SEC column (9).

18. It is advisable to compare monodispersity with a membrane protein sample that has previously been purified and solubilized in your selected detergent. If none of the selected homologues are highly produced and monodisperse in a mild detergent, and the addition of cholesterol and/or ligand do not improve monodispersity, consider constructing further mutants or screening more homologues. Using this strategy we and others have purified a number of mammalian GPCRs, transporters, channels, and enzymes (5, 6, 10, 11).

Acknowledgments

This work was supported by the Royal Society (United Kingdom) through a University Research Fellowship to DD and by a Basic Science Research Program grant through the National Research Foundation of Korea (NRF) funded by the Ministry of Education, Science and Technology (NRF0409-20100093) to HK.

References

1. Drew DE, von Heijne G, Nordlund P, de Gier JW (2001) Green fluorescent protein as an indicator to monitor membrane protein over-expression in *Escherichia coli*. FEBS Lett 507:220–224

2. Drew D, Slotboom DJ, Friso G, Reda T, Genevaux P, Rapp M, Meindl-Beinker NM, Lambert W, Lerch M, Daley DO, Van Wijk KJ, Hirst J, Kunji E, De Gier JW (2005) A scalable GFP-based pipeline for membrane protein overexpression screening and purification. Protein Sci 14:2011–2017

3. Drew D, Lerch M, Kunji E, Slotboom DJ, de Gier JW (2006) Optimization of membrane protein overexpression and purification using GFP fusions. Nat Methods 3:303–313

4. Drew D, Newstead S, Sonoda Y, Kim H, von Heijne G, Iwata S (2008) GFP-based optimization scheme for the overexpression and purification of eukaryotic membrane proteins in *Saccharomyces cerevisiae*. Nat Protoc 3:784–798

5. Sonoda Y, Cameron A, Newstead S, Omote H, Moriyama Y, Kasahara M, Iwata S, Drew D (2010) Tricks of the trade used to accelerate high-resolution structure determination of membrane proteins. FEBS Lett 584:2539–2547

6. Warne T, Serrano-Vega MJ, Tate CG, Schertler GF (2009) Development and crystallization of a minimal thermostabilised G protein-coupled receptor. Protein Expr Purif 65:204–213

7. Bowie JU (2001) Stabilizing membrane proteins. Curr Opin Struct Biol 11:397–402

8. Newstead S, Kim H, von Heijne G, Iwata S, Drew D (2007) High-throughput fluorescent-based optimization of eukaryotic membrane protein overexpression and purification in *Saccharomyces cerevisiae*. Proc Natl Acad Sci USA 104:13936–13941

9. Kawate T, Gouaux E (2006) Fluorescence-detection size-exclusion chromatography for precrystallization screening of integral membrane proteins. Structure 14:673–681

10. Iizasa E, Mitsutomi M, Nagano Y (2010) Direct binding of a plant LysM receptor-like kinase, LysM RLK1/CERK1, to chitin *in vitro*. J Biol Chem 285:2996–3004

11. Sugawara T, Ito K, Shiroishi M, Tokuda N, Asada H, Yurugi-Kobayashi T, Shimamura T, Misaka T, Nomura N, Murata T, Abe K, Iwata S, Kobayashi T (2009) Fluorescence-based optimization of human bitter taste receptor expression in *Saccharomyces cerevisiae*. Biochem Biophys Res Commun 382:704–710

Chapter 9

The Effect of Antifoam Addition on Protein Production Yields

Sarah J. Routledge and Roslyn M. Bill

Abstract

Pichia pastoris is a widely used host for recombinant protein production. The foaming associated with culturing it on a large scale is commonly prevented by the addition of chemical antifoaming agents or "antifoams." Unexpectedly, the addition of a range of antifoams to both shake flask and bioreactor cultures of *P. pastoris* has been shown to alter the total yield of the recombinant protein being produced. Possible explanations for this are that the presence of the antifoam increases the total amount of protein being produced and secreted per cell or that it increases the density of the culture. Antifoaming agents may therefore have specific effects on the growth and yield characteristics of recombinant cultures, in addition to their primary action as de-foamers.

Key words: Antifoam, Protein, Foam destruction yield, k_La, Viability

1. Introduction

When producing recombinant proteins on a large scale, growth in bioreactors is often the most convenient option. One problem with such cultures, which are typically intensely aerated and stirred, is the formation of foam. Foaming can lead to reduced process productivity since the bursting of bubbles may damage cells and proteins (1) and result in loss of sterility if the foam escapes the bioreactor (2). To prevent the formation of foam, thermal methods (3), mechanical agitation, ultrasound, or, most often, addition of chemical antifoaming agents or "antifoams" (2) is employed.

Chemical antifoams can act as surfactants and generally consist of several components. There are many different varieties which can usually be grouped as hydrophobic solids dispersed in carrier oil, aqueous or water-based suspensions or emulsions, liquid single-component antifoams and solid antifoams (3, 4). There are thought

Roslyn M. Bill (ed.), *Recombinant Protein Production in Yeast: Methods and Protocols*, Methods in Molecular Biology, vol. 866, DOI 10.1007/978-1-61779-770-5_9, © Springer Science+Business Media, LLC 2012

to be several mechanisms of foam dispersion by chemical antifoams: bridging-dewetting, spreading fluid entrainment, and bridging-stretching (5), but the precise details are not well understood.

It is clear that antifoams are often added to bioreactors without considering the effects on the bioprocess, the cells or the recombinant proteins being produced. This is surprising as antifoams are known to affect the $k_L a$ or oxygen transfer rate of a system (6–14), the ability of cells to secrete protein and can even be used at high concentrations to boost the yield of a recombinant protein in shake flasks (15). However, different types of antifoam at varying concentrations can have different effects. These properties should be taken into account when choosing the most suitable antifoam for a particular process.

1.1. Antifoams

A wide variety of antifoams is available, with different compositions and properties. One consideration when deciding on the type of antifoam to be used is whether it is suitable for a particular process. Certain antifoams may not be suitable for some applications, such as producing proteins for drug development, as not all are FDA approved. Some antifoams may also affect downstream processing, such as silicone-containing antifoams, which can coat equipment. Several antifoams should therefore be tested to determine which is the most appropriate for the process.

1.2. Bartsch Test

When selecting an antifoam, its primary function as a de-foamer should be evaluated. A simple Bartsch test (16, 17) can be performed to asses foam-destruction properties. This involves shaking the medium to be used in the process to induce foam formation. When this is done in the presence of the antifoam of interest, it is possible to assess the ability of the agent to destroy foam over time.

1.3. $k_L a$

The $k_L a$, or volumetric mass oxygen transfer coefficient, is a measure of how much oxygen is transferred into the medium over a certain amount of time (14). The $k_L a$ of a system can be influenced by several factors such as the properties of the medium including its viscosity, the presence of organisms and their by-products. Additions to the medium such as antifoams also have an effect (13, 14). It has been observed that low concentrations of antifoam can reduce the $k_L a$ but at higher concentrations the $k_L a$ may rise (10, 12). To ensure optimum oxygen transfer within a system, the effect of differing concentrations of the antifoam to be used should be assessed.

1.4. Influence of Antifoams on Protein Yield

Antifoams added to shake flask cultures at higher concentrations than normally used in bioreactors have been shown to increase the yield of recombinant green fluorescent protein (GFP) (15). Two groups of antifoams were identified in the study, depending on their mode of action: one that resulted in improvements in biomass yields of the cultures and one that enhanced protein production or secretion. This finding may provide a simple method to increase productivity in recombinant protein production experiments. It also provides much needed insight into how antifoams interact with yeast host cells.

1.5. Effect of Antifoams on Cell Viability

A check should be done to ensure antifoams at high concentrations are not detrimental to the viability of cells. Flow cytometry can be performed using shake flask samples stained with propidium iodide. In this assay, dead cells are stained red.

2. Materials

2.1. Antifoams

Antifoams from different groups can be selected depending on process requirements, e.g.,

- Antifoam A (Sigma), 30% emulsion of silicone polymer.
- Antifoam C (Sigma), 30% emulsion of silicone polymer.
- Antifoam J673A (Struktol), an alkoxylated fatty acid ester on a vegetable base.
- Antifoam P2000 (Fluka), a polypropylene glycol.
- Antifoam SB2121 (Struktol), a polyalkylene glycol.

2.2. Bartsch Test

1. Culture medium required for the process, e.g., BMMY medium for *Pichia pastoris* cultures is composed of 1% yeast extract, 2% peptone, 100 mM potassium phosphate pH 6.0, 1.34% YNB, 4×10^{-5}% biotin, 0.5% methanol. Dissolve 10 g yeast extract and 20 g peptone in 700 mL water and autoclave at 121°C for 20 min. After cooling to room temperature, add 100 mL 1 M potassium phosphate buffer pH 6.0, 100 mL 10× YNB, 2 mL biotin, and 100 mL 10× methanol. Store at 4°C with a shelf life of approximately 2 months.

2. YNB: dissolve 134 g yeast nitrogen base with ammonium sulfate and without amino acids in water to a total volume of 1 L and filter sterilize. Store at 4°C with a shelf life of approximately 1 year.

3. 500× biotin (0.02%): dissolve 20 mg biotin in water to a total volume of 100 mL and filter sterilize. Store at 4°C with a shelf life of approximately 1 year.

4. 10× methanol (5%): mix 5 mL methanol with 95 mL water and filter sterilize. Store at 4°C with a shelf life of approximately 2 months.

5. 1 M potassium phosphate buffer pH 6.0: Combine 132 mL of 1 M K_2HPO_4 with 868 mL KH_2PO_4 and set the pH to 6.0 using a pH meter and phosphoric acid. Autoclave the solution and store at room temperature with a shelf life of over 1 year.

6. 300 mL graduated glass measuring cylinder.

7. Parafilm to seal the measuring cylinder.

8. Stopwatch.

2.3. $k_L a$ Measurement

1. Data logging equipment to record dissolved oxygen (DO) measurements rapidly, such as a Picolog ADC-16 (Applikon) which records data every 300 ms from the DO probe. This can be used with PicoLog software to display a plot of the output. The DO is recorded in mV by this equipment and can be converted to % DO once the mV readings at 0% and 100% DO have been determined.

2. The medium used for $k_L a$ measurements should be the same as for a full bioreactor run, e.g., BMMY (see Subheading 2.2).

3. A bioreactor with control systems and equipment usually used for a full bioprocess, including a DO probe.

2.4. Shake Flask Cultures of Recombinant GFP

1. An agar plate with colonies of the organism of interest, e.g. *P. pastoris* X33 GFP grown on YPD agar composed of 1% yeast extract, 2% peptone, 2% dextrose (glucose), 2% agar. Dissolve 20 g peptone and 10 g yeast extract in water to a total volume of 900 mL and add 20 g agar. Autoclave the solution at 121°C for 20 min and cool to room temperature before adding 100 mL 10× glucose and pouring plates which should be stored at 4°C.

2. 10× glucose: dissolve 200 g glucose in water to a total volume of 1 L, autoclave at 121°C for 20 min, and then cool to room temperature. Store at 4°C.

3. Medium to be used to set up a starter culture, e.g. BMGY medium composed of 1% yeast extract, 2% peptone, 100 mM potassium phosphate pH 6.0, 1.34% YNB, 4×10^{-5}% biotin, 1% glycerol. Dissolve 10 g yeast extract and 20 g peptone in water to a total volume of 700 mL. Autoclave at 121°C for 20 min. After cooling to room temperature, add 100 mL 1 M potassium phosphate buffer pH 6.0, 100 mL 10× YNB, 2 mL biotin, and 100 mL 10× glycerol. The final volume should be adjusted to 1 L. Store at 4°C with a shelf life of approximately 2 months.

4. BMMY medium (see Subheading 2.2).

5. One 250-mL baffled shake flask and 18 × 100-mL non-baffled shake flasks for antifoam evaluations, all autoclaved.

6. Spectrophotometer to determine the optical density of cultures.

7. 100% methanol for inducing protein production.

2.5. Assessment of Cell Viability by Flow Cytometry

1. PBS pH 7.0: dissolve one 10× PBS tablet in 1 L water.

2. Light microscope.

3. Hemocytometer.

4. Flow cytometer, e.g., Beckman Coulter (High Wycombe, UK) flow cytometer with 488 nm excitation from an argon-ion laser at 15 mW.

5. Propidium iodide (1 mg/mL) in water.

6. WinMDI software for data analysis.

3. Methods

3.1. Bartsch Test

1. This method is adapted from Denkov et al. (16).

2. Fill a 300-mL graduated glass measuring cylinder with 100 mL medium.

3. Pipette the antifoam to be tested into the medium to 0.01% v/v, i.e., 10 µL (see Notes 1 and 2).

4. Seal the cylinder using parafilm and shake up and down ten times at ambient temperature to induce foam formation.

5. Record the total volume of the system (the medium and foam combined), and the volume of the medium alone every 30 s for 15 min (see Note 3).

6. Determine the activity by subtracting the volume of the medium from the total volume.

7. Compare the effects of different antifoams to a control with no added antifoam to select the most efficient agent. An example is shown in Fig. 1.

3.2. Measurement of $k_L a$

1. The dynamic method of $k_L a$ measurement is based on the method outlined by Bandyopadhyay and Humphrey (18). Here the experimental set-up is a 3-L glass bioreactor (Applikon Biotechnology) containing BMMY medium in the absence of cells.

2. A working volume of 1 L medium is used and the bioreactor set up for conditions typical of the required bioprocess, e.g. 1.0 L/min compressed air (60% O_2), pH 6.0, 30°C, and 700 rpm (see Note 4).

3. Calibrate the DO probe by measuring the DO in the bioreactor at these settings for 100% (60% O_2 in compressed air will achieve 100% DO), then flush with nitrogen gas instead of air to obtain a 0% DO reading.

4. $k_L a$ measurements are carried out by starting at 100% DO and flushing with nitrogen until the DO drops to 0%. Supply the system with air and the DO gradually rises to 100%, whereupon it is again flushed with nitrogen to reduce it to 0% before reconnecting the air. The time points are recorded by the data logger and the DO can be plotted versus time, as shown in Fig. 2.

5. If several concentrations of antifoam are to be tested, add them in a stepwise manner to the bioreactor once the DO is at 100%. Follow each addition by flushing with nitrogen, as in step 4.

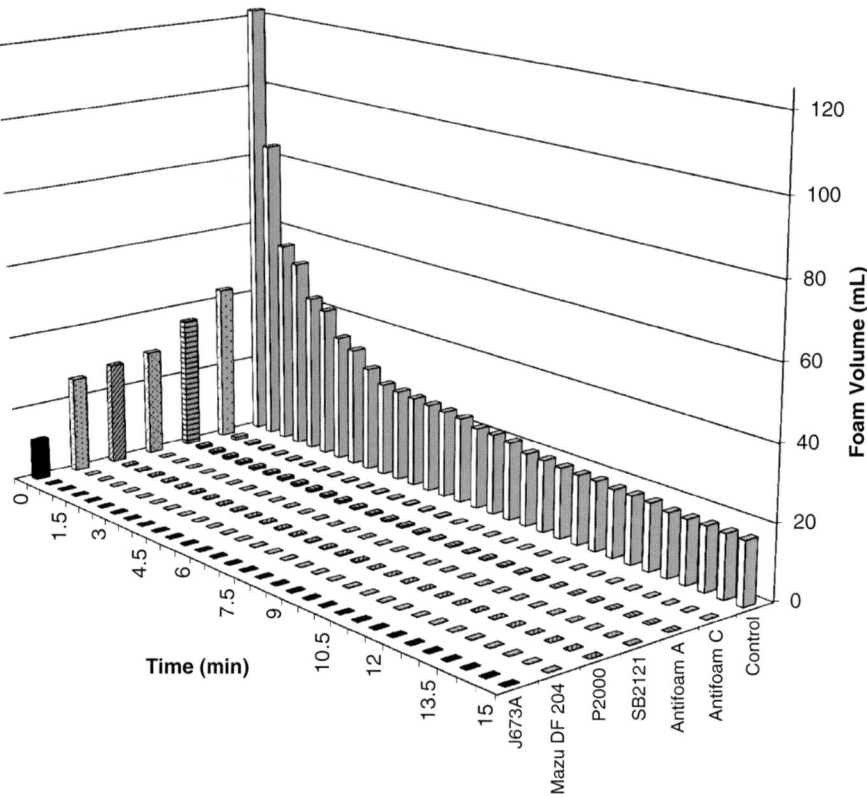

Fig. 1. A Bartsch test of foam volume over time for various antifoams in BMMY medium. Foam volume over time was recorded for 0.01% v/v of each antifoam in BMMY medium where $n = 5$. All antifoams were effective at foam destruction compared with the control and most foam was destroyed within 1 min.

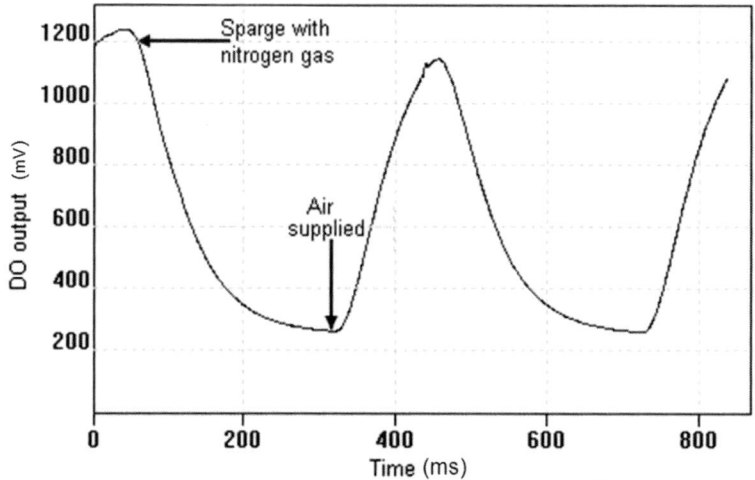

Fig. 2. DO over time for a 3-L glass bioreactor containing 1 L BMMY medium. The bioreactor was flushed with nitrogen to reduce the DO, followed by air. The DO was recorded in mV with a data logger and PicoLog software.

6. The data generated for the upward slope of the plot are used to calculate the k_La with the following equation:

$$k_La(t_2 - t_1) = \ln\left(\frac{c_{1,\infty} - c_{1,t1}}{c_{1,\infty} - c_{1,t2}}\right)$$

where t_1 and t_2 are consecutively logged time points, $c_{1,\infty}$ is the oxygen saturation concentration, and c_1 is the oxygen concentration at each time point. An example calculation can be found in Subheading 4 (see Note 5).

7. $c_{1,\infty}$ is calculated using the following constants:

- XO_2 (fraction of O_2 in air) $= 0.2095$
- MO_2 (O_2 partition coefficient) $= 30$
- R (gas constant) $= 8.3144$ J/K/mol

And the following variables:

- T (temperature) $= 30°C$ or 303 K
- P (approximate pressure in a 3-L glass bioreactor) $= 0.1$ barg or 1.1 barg
- C (concentration of all gas in the head space)
- n (number of moles of gas in the head space)
- V (volume of gas in the head space)

8. Combining ($C = n/V$) and the universal gas law ($PV = nRT$) gives $C = P/RT$, allowing the concentration of air in the gas phase in the head space of the vessel, $CgAir$, to be calculated.

9. The concentration of O_2 in the gas phase, CgO_2, can then be calculated by multiplying the concentration of gas in air by XO_2.

10. Dividing the gas concentration of O_2 by MO_2 gives the maximum liquid oxygen saturation concentration, $c_{1,\infty}$ at 100% DO.

11. DO percentage values at a particular time point are converted to oxygen concentrations by dividing by 100 and multiplying by $c_{1,\infty}$.

3.3. Analyzing the Influence of Antifoams on the Yield of GFP in Shake Flasks

1. Pick a single colony of *P. pastoris* X33 GFP from a YPD plate and inoculate a 250-mL baffled shake flask containing 50 mL of BMGY.

2. Incubate at 30°C, 220 rpm overnight.

3. Measure the optical density at 595 nm using a spectrophotometer. A typical value would be 20.

4. Calculate the volume of culture necessary to inoculate the required volume of BMMY to achieve a final OD_{595} of 1. In this example, the required volume of BMMY is 360 mL.

5. Centrifuge this volume of culture at $5,530 \times g$ and discard the supernatant.

6. Resuspend the pellet in 360 mL BMMY and mix.

7. Dispense 20 mL culture into each of the 18×100 mL non-baffled shake flasks.

8. Incubate the flasks with the desired concentration of antifoam at 30°C, 220 rpm. A suggested experimental set-up is triplicate flasks for antifoams at 0%, 0.2%, 0.4%, 0.6%, 0.8%, and 1.0%.

9. Record the optical density of the samples at 0 h, 24 h, and 48 h.

10. After 24 h add 100% sterile methanol to 1% v/v, i.e., 200 μL to 20 mL, and continue the incubation.

11. At 48 h harvest the cultures.

12. Analyze the protein content. For secreted GFP, centrifuge 1 mL samples at $18,625 \times g$ for 10 min and separate the pellet and supernatant. Analyze the supernatant samples on a fluorescence plate reader (Fig. 3).

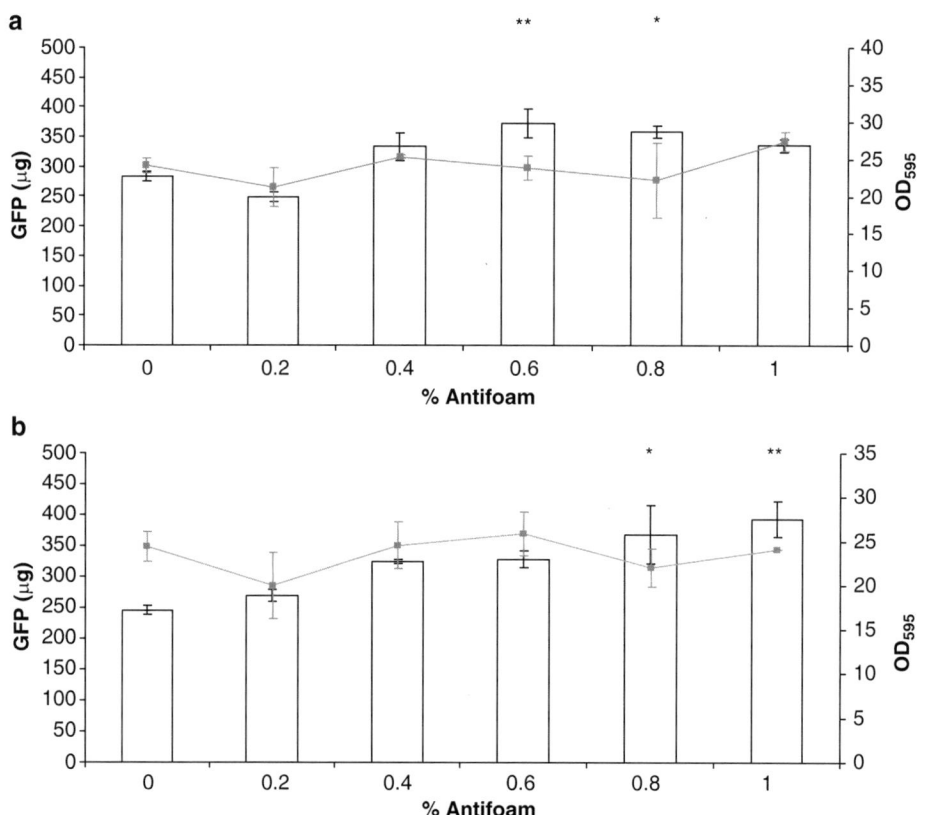

Fig. 3. Effect of antifoams on the total yield of GFP produced by 20 mL recombinant *P. pastoris* X33 over 48 h. The yield of secreted GFP was influenced by antifoam A (**a**) and J673A (**b**). The *horizontal line* represents the average optical density for each concentration of antifoam and $n=9$ for each point. The data were analyzed using a one-way ANOVA ($P < 0.001$) and a Dunnett's multiple comparison test where $* = P \leq 0.05$, $** = P \leq 0.01$. Each of the antifoams increased the yield at higher concentrations than would normally be used in bioreactors.

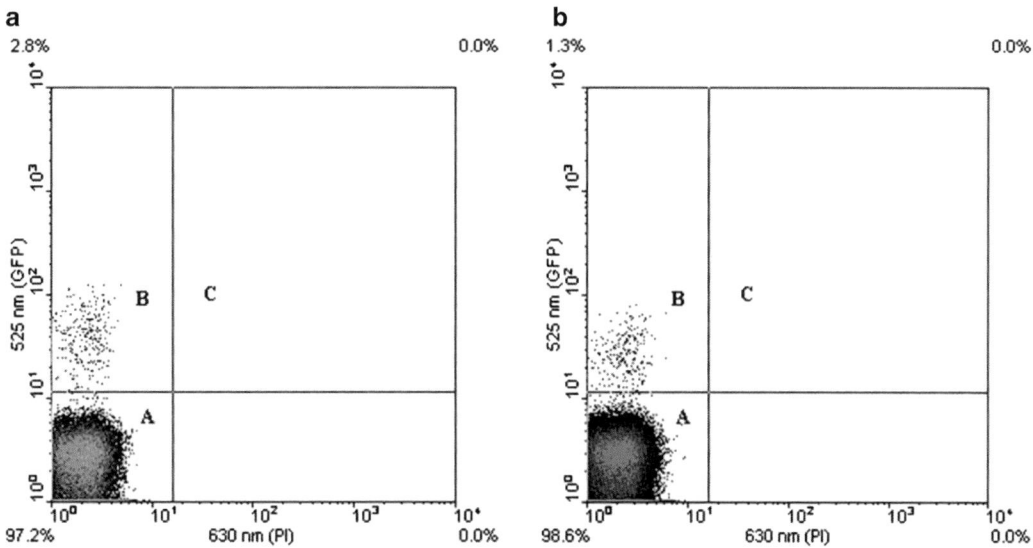

Fig. 4. Viable cells without antifoam (**a**) and viable cells with 0.6% antifoam A (**b**). Population A is made of events that are related to electronic and particulate noise and are not cells. Population B are cells showing enhanced green fluorescence due to GFP and quadrant C is where dead cells stained red with propidium iodide (PI) would be observed. Antifoam A did not adversely affect the viability of the cells.

3.4. Analyzing the Effect of Antifoams on the Viability of Cells

1. Dilute samples from Subheading 3.3 step 11 using PBS to obtain a concentration of 10^6 to 10^7 cells/mL. Use a hemocytometer and a light microscope.

2. Stain 2 mL of each sample by adding 10 μL propidium iodide.

3. Analyze the samples using a flow cytometer. Propidium iodide fluorescence is measured at 630 nm and GFP fluorescence at 525 nm.

4. WinMDI software can be used to plot the data, as shown in Fig. 4.

4. Notes

1. Antifoams may be sterilized prior to use by autoclaving at 121°C for 20 min, although this is not necessary.

2. Some antifoams are viscous and sticky and may take several seconds to be drawn into a pipette tip of the required volume. Additionally, they may stick to the inside of the pipette tip and therefore must be dispensed fully by repeatedly drawing medium into the tip and pipetting it out.

3. Foam may reach different heights around the sides of the cylinder, therefore the highest level of foam should be recorded.

4. A lower impeller speed may be required before adding antifoam as the foam level may become too high at higher speeds depending upon the type of medium.

5. An example of a $k_L a$ calculation using the constants and variables from Subheading 3.2 step 7 is given here:

   (a) Calculate the concentration of air in the gas phase at 100% DO:

   - $C = P/RT$
   - Therefore $CgAir = 1 \times 10^4 / (8.3145 \times 303)$
     $$= 3.97 \text{ mol/m}^3$$

   (b) Using the concentration of air in the gas phase, the concendration of O_2 present in the gas phase is calculated:

   - $CgO_2 = XO_2 \times CgAir$
     $$= 0.2095 \times 3.97$$
     $$= 0.83 \text{ mol/m}^3$$

   (c) Convert O_2 in the gas phase to O_2 in the liquid phase for the maximum liquid saturation concentration:

   - $c_{1,\infty} = CgO_2 / MO_2$
     $$= 0.83/30$$
     $$= 0.028 \text{ mol/m}^3 \text{ at } 100\% \text{ DO}$$

   (d) Convert DO values at a given time point to oxygen saturation concentrations, C_1, by dividing by 100 and multiplying by $c_{1,\infty}$:

   - e.g., oxygen saturation concentration at 72% DO
   - C_1 at 72% DO $= (72/100) \times 0.028$
     $$= 0.02 \text{ mol/m}^3$$

   (e) Substitute the C_1 values calculated at particular time points into the equation in Subheading 3.2 step 6 to calculate the $k_L a$.

References

1. Holmes WJ, Smith R, Bill RM (2006) Evaluation of antifoams in the expression of a recombinant F_C fusion protein in shake flask cultures of *Saccharomyces cerevisiae*. Microb Cell Fact 5:30

2. Varley J, Brown A, Boyd R, Dodd P, Gallagher S (2004) Dynamic multipoint measurement of foam behaviour for a continuous fermentation over a range of key process variables. Biochem Eng J 20:61–72

3. Höefer R, Jost F, Schwuger MJ, Scharf R, Geke J, Kresse J, Lingman H, Veitenhansl R, Erwied W (2000) Uilmann's Encyclopedia of Industrial Chemistry. Wiley-VCM

4. Joshi K, Jeelani S, Blickenstorfer C, Naegeli I, Windhab E (2005) Influence of fatty alcohol antifoam suspensions on foam stability. Colloids Surf A 263:239–249

5. Denkov ND, Krastanka M, Christova C, Hadjiiski A, Cooper P (2000) Mechanisms of action of mixed solid–liquid antifoams: 3. Exhaustion and reactivation. Langmuir 21:8163–8619

6. Al-Masry W (1999) Effects of antifoam and scale-up on operation of bioreactors. Chem Eng Process 38:197–201

7. Arjunwadkar SJ, Sarvanan K, Kulkarni PR, Pandit AB (1998) Gas–liquid mass transfer in dual impeller bioreactor. Biochem Eng J 1:99–106

8. Calik P, Ileri N, Erdinc BI, Aydogan N, Argun M (2005) Novel antifoam for fermentation processes: fluorocarbon-hydrocarbon hybrid unsymmetrical bolaform surfactant. Langmuir 21:8613–8619

9. Koch V, Rüffer H, Schügerl K, Innertsberger E, Menzel H, Weis J (1995) Effect of antifoam agents on the medium and microbial cell properties and process performance in small and large reactors. Process Biochem 30:435–446

10. Morao A, Maia C, Fonseca M, Vasconcelos J, Alves S (1999) Effect of antifoam addition in gas–liquid mass transfer in stirred fermenters. Bioprocess Eng 20:165–172

11. Koide K, Yamazoe S, Harada S (1985) Effects of surface-active substances on gas hold up and gas–liquid mass transfer in bubble column. J Chem Eng Jpn 18:287–292

12. Liu H-S, Chiung W-C, Wang Y-C (1994) Effect of lard oil and castor oil on oxygen transfer in an agitated fermentor. Biotechnol Tech 8:17–20

13. Yagi H, Yoshida F (1974) Oxygen absorption in fermenters – effects of surfactants, antifoaming agents and sterilized cells. J Ferment Technol 52:905–916

14. Stanbury PF, Whittaker A, Hall SJ (1995) Principles of fermentation technology, 2nd edn. Butterworth Heinemann, Oxford

15. Routledge SJ, Hewitt CJ, Bora N, Bill RM (2011) Antifoam addition to shake flask cultures of recombinant *Pichia pastoris* increases yield. Microb Cell Fact 10:17

16. Denkov ND, Tcholakova S, Marinova KG, Hadjiiski A (2002) Role of oil spreading for the efficiency of mixed oil–solid antifoams. Langmuir 18:5810–5817

17. Bartsch O (1924) Über Schaumsysteme. Fortschrittsberichte über Kolloide und Polymere 20:1–49

18. Bandyopadhyay P, Humphrey AE (1967) Dynamic measurement of the volumetric oxygen transfer coefficient in fermentation systems. Biotechnol Bioeng 9:533–544

Chapter 10

Setting Up a Bioreactor for Recombinant Protein Production in Yeast

Sarah J. Routledge and Michelle Clare

Abstract

Scale-up from shake flasks to bioreactors allows for the more reproducible, high-yielding production of recombinant proteins in yeast. The ability to control growth conditions through real-time monitoring facilitates further optimization of the process. The setup of a 3-L stirred-tank bioreactor for such an application is described.

Key words: Bioreactor, Scale-up, *Pichia pastoris*, *Saccharomyces cerevisiae*

1. Introduction

Previous chapters have described the use of shake flasks for culture growth. Shake flasks are useful in screening for high-yielding clones and maintaining basic cultivation parameters, such as temperature. However, high yields of protein are often not achieved. Despite recent developments allowing pH and dissolved oxygen (DO) monitoring in shake flasks, there remain restrictions in volume and oxygen transfer rates, as well as an inability to adequately control pH and DO in the flask format. In contrast, scale-up to bioreactors provides tight control and allows online monitoring of culture conditions, thereby facilitating the development of effective feeding strategies and further optimization of a given bioprocess to produce much greater quantities of protein (1).

Recombinant protein production experiments in *Pichia pastoris* and *Saccharomyces cerevisiae* are most commonly optimized in bioreactors by varying feeding and induction strategies (2, 3). For *P. pastoris*, induction with mixed feeds of glycerol and methanol or glycerol and sorbitol has been investigated (4, 5). It has also been demonstrated that strict maintenance of the culture pH (6) and variation of

Roslyn M. Bill (ed.), *Recombinant Protein Production in Yeast: Methods and Protocols*, Methods in Molecular Biology, vol. 866, DOI 10.1007/978-1-61779-770-5_10, © Springer Science+Business Media, LLC 2012

the temperature during the induction step are important factors in increasing protein yields (7), both of which are simple to accomplish using a bioreactor. The optimum pH, temperature, and DO required can vary with each strain and the protein being produced. For example, Jamshad and colleagues found that the growth temperature for the production of the human tetraspanin, CD81, in *P. pastoris* was 30°C (8) while in another study Dragosits and colleagues demonstrated that decreasing temperature from 30 to 20°C was optimal for the production of the antibody fragment, Fab 3 H6 (9). Clearly, a careful analysis of how the culture responds to the growth conditions is vital in achieving high yields of protein.

1.1. Setting Up a Bioreactor for *Pichia pastoris*

A commonly used bioreactor for recombinant protein production is the stirred-tank reactor with agitation provided by impellors (10). This chapter describes how to set up and run a 3-L stirred-tank autoclavable bioreactor using equipment from Applikon Biotechnology for *P. pastoris* cultures. Once an agar plate with the desired colonies has been grown, the process can be completed within a week or less, depending upon the length of the induction phase.

Day 1 consists of setting up a seed culture and preparing the bioreactor. Day 2 involves setting the process parameters and inoculating the bioreactor. The batch phase continues until day 3; after observing a DO spike indicating that the glycerol present in the growth medium has been metabolized (20–24 h after inoculation), a fed-batch phase is begun to increase biomass. After 4 h of fed batch, the culture is starved for an hour to ensure that all remaining glycerol is utilized. Induction with methanol is then performed to initiate recombinant protein production and can continue to around 96 h post inoculation, making 6 days total.

2. Materials

2.1. Seed Culture Materials

1. Prepare an agar plate with colonies of the organism of interest, e.g., *P. pastoris* X33 GFP grown on YPD agar (11) composed of 1% yeast extract, 2% peptone, 2% dextrose, and 2% agar. 20 g peptone and 10 g yeast extract are dissolved in water to a total volume of 900 mL and 20 g agar is added. The solution is autoclaved at 121°C for 20 min and cooled to room temperature before adding 100 mL 10× dextrose and pouring plates which are stored at 4°C.

2. 10× dextrose: Dissolve 200 g glucose in water to a total of 1 L, autoclave at 121°C for 20 min, and then cool to room temperature. The solution is stored at 4°C.

3. Prepare medium for the seed culture, e.g. BMGY (11) medium composed of 1% yeast extract, 2% peptone, 100 mM potassium phosphate, pH 6.0, 1.34% YNB, 4×10^{-5}% biotin, and 1% glycerol. 10 g yeast extract and 20 g peptone are dissolved in water to a total volume of 700 mL. The solution is autoclaved at 121°C for 20 min. After cooling to room temperature, 100 mL 1 M potassium phosphate buffer, pH 6.0, 100 mL 10× YNB, 2 mL biotin, and 100 mL 10× glycerol are added. Store at 4°C with a shelf life of approximately 2 months.

4. Autoclave one 250-mL baffled shake flask.

5. Spectrophotometer to determine the optical density of cultures.

2.2. Bioreactor Materials

1. Prepare 1 L basal salts medium (BSM) (12) by dissolving 26.7 mL 85% phosphoric acid, 0.93 g calcium sulfate, 18.2 g potassium sulfate, 14.9 g magnesium sulfate heptahydrate, 4.13 g potassium hydroxide, and 40.0 g glycerol in water to a total of 1 L. The solution is autoclaved after being added to the bioreactor.

2. Prepare 1 L PTM_1 trace salts (12) by dissolving 6.0 g cupric sulfate pentahydrate, 0.08 g sodium iodide, 3.0 g manganese sulfate monohydrate, 0.2 g sodium molybdate dihydrate, 0.02 g boric acid, 0.5 g cobalt chloride, 20.0 g zinc chloride, 65.0 g ferrous sulfate heptahydrate, 0.2 g biotin, and 5.0 mL sulfuric acid in water to a total of 1 L. The solution is filter sterilized and stored at room temperature.

3. Prepare 0.5 L 50% (w/v) glycerol feed (12) by dissolving 250 g glycerol in water to a total of 0.5 L. Dispense into a feed bottle with tubing of appropriate diameter for the intended pump, clamp the line, and cover the end in foil. Attach a filter to the remaining line. Autoclave at 121°C for 20 min, and then add 6 mL PTM_1 trace salts.

4. Prepare 1 L 20% (v/v) methanol feed (4) by dispensing 800 mL water into a feed bottle with tubing of appropriate diameter for the intended pump, clamping the line and covering the end in foil. Attach a filter to the remaining line. Autoclave the bottle at 121°C for 20 min and then add 200 mL 100% methanol and 12 mL PTM_1 trace salts aseptically in a laminar flow cabinet.

2.3. Other Solutions

1. Prepare a 0.5-L glass liquid addition bottle containing 50% (v/v) phosphoric acid and a 0.5-L glass liquid addition bottle of 28% ammonium hydroxide for pH control. PharMed tubing may be used and filters attached to the caps. Clamp the tubing until needed and clearly label the bottles.

2. Prepare 20 mL 50% (w/v) glycerol to aid heat conductivity to the temperature probe by dissolving 10 g glycerol and making

up to 20 mL with water. Sterilization is not necessary as the solution does not come into contact with the medium inside the bioreactor.

3. Prepare a wash bottle of 70% ethanol by mixing 350 mL ethanol with 150 mL water.

2.4. Bioreactor Equipment

The parts listed are for use with a 3 L stirred tank autoclavable bioreactor (Applikon) using an ADI 1010 controller and BioXpert version 2 software.

1. Jacketed 3 L glass vessel (Applikon).
2. Head plate to fit a 3 L vessel (Applikon).
3. BioXpert software (Applikon) installed onto a PC.
4. Thermo circulator, ADI 1018.
5. Tandem off-gas analyser.
6. Gas supply unit, ADI 1026 (Applikon).
7. 60% oxygen: 40% nitrogen gas cylinder (BOC).
8. DO probe (Applikon).
9. DO probe electrolyte and membranes (Applikon).
10. pH probe (Applikon).
11. pH 4 and pH 7 buffers (Fisher).
12. Optek probe (optical density).
13. Optek controller.
14. 250 mL glass sample bottle (Fisher).
15. Air compressor (Bambi, 75/150).
16. Recirculating chiller (Grant, LTL1).
17. Peristaltic pump (×2) for acid and base addition (Easyload Masterflex).
18. Peristaltic pump for feeds (Masterflex, C/L).
19. Filters (Sartorius, midistart 2000 0.2 μm PTFE).
20. Silicon tubing (Fisher).
21. Tubing connectors and Y connectors (Fisher).
22. Clamps (Fisher).
23. Needles (Fisher).
24. Plastic syringes (Fisher, 5, 20, and 50 mL).
25. 0.5 L glass liquid addition bottles (×3; Applikon).
26. 1 L glass liquid addition bottle (Applikon).
27. Foam bungs (Fisher).
28. Tin foil (Fisher).
29. 20 mL sample tubes (Fisher).
30. Spectrophotometer.

3. Methods

3.1. Preparation of the Seed Culture

1. A single colony of *P. pastoris* from a YPD plate is used to inoculate a 250-mL baffled shake flask containing 50 mL BMGY. This should be done the day before starting the bioreactor culture.

2. The flask is incubated at 30°C, 220 rpm agitation, overnight.

3.2. Preparation of the DO Probe

1. Check that the DO probe is clean and remove the protective cap.

2. Ensure that the O-rings are undamaged.

3. Check that the membrane is clean and if not, clean using toothpaste and a toothbrush. Ensure that the membrane is not damaged or replace if this is the case.

4. Unscrew the membrane module and using a magnifying glass, make sure that the bulb at the base of the probe is clean and free from cracks. Clean by using tissue and toothpaste, gently twisting the bulb against the tissue.

5. Ensure that there is 1 mL electrolyte solution in the bottom of the membrane module and screw back onto the probe.

3.3. Preparation and Calibration of the pH Probe and Pumps

1. Check that the pH probe is clean and remove the protective cap.

2. Ensure that the O-rings are undamaged.

3. Remove the cap at the top of the probe and connect to the pH cable on the ADI 1010. Rinse the probe using deionized water and dry, and then place the probe into pH 4 buffer.

4. Turn on the ADI 1010 and in the main menu screen press the calibrate button (Calib).

5. Select calibrate using the dial and press the calibrate button.

6. Enter the temperature of the buffer and press calibrate.

7. Using the dial, enter the buffer pH and wait for the reading to stabilize which can take a few minutes. When the reading is stable, press the calibrate button.

8. Rinse the probe and dry before placing in pH 7 buffer. Using the dial, enter the buffer pH and wait for the reading to stabilize before pressing the calibrate button.

9. The calibration slope and offset are displayed once calibration is complete. Ensure that the values for the slope are within the manufacturer's specification, in this case 0.95–1.05, and for the offset they are $< \pm 0.3$ (13).

10. Press the set point button (Setp) twice to return to the main menu screen.

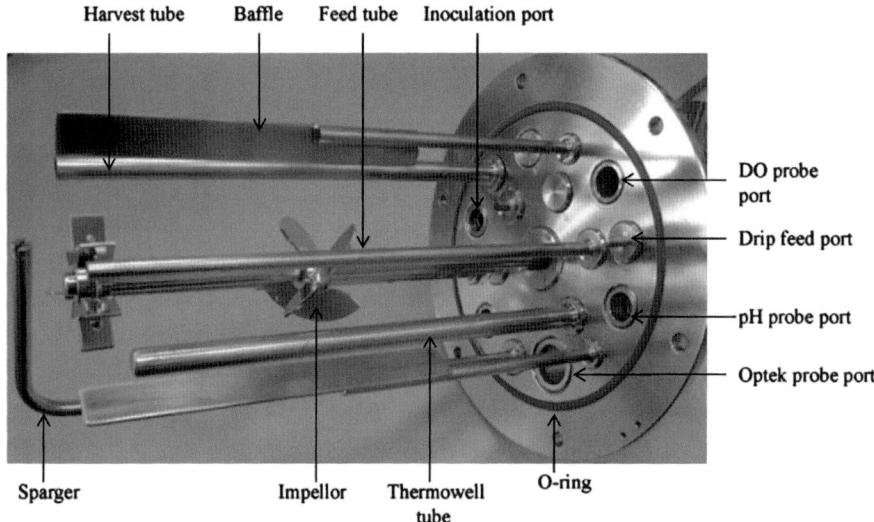

Fig. 1. A photograph showing the head plate layout for an Applikon 3 L stirred tank autoclavable bioreactor in the laboratory of Dr Roslyn Bill, Aston University, UK. The various features are marked.

11. Disconnect the pH probe from the ADI 1010 and replace the cap on the top of the probe.

12. Calibrate the pump flow rate with your particular tubing.

3.4. Preparation of the Bioreactor

1. Ensure that the bioreactor head plate and glass vessel are clean by washing with warm water. Run warm water through the sample ports and sparger to remove any trapped particles. Check that the septum is clean and without any holes.

2. Check that the baffles and impellors are tightly attached to the head plate as they could damage probes if they become loose during the run (see Fig. 1).

3. Check that the O-rings on the base of the head plate are in place and undamaged.

4. Make 1 L BSM and pour into the glass vessel.

5. Place the head plate onto the glass vessel and attach using the bolts (see Fig. 2). Check that all bolts are tight.

6. Ensure that silicon tubes are attached to each of the ports on top of the head plate. Check that they are free from holes and that the tubing is not worn or old. Clamp the ends of each tube and cover the ends in foil. For the feed lines, use the ports that allow solutions to drip into the vessel, thereby allowing a visual check to be made when feeds are started.

7. Place the pH probe after calibrating into the appropriate port and screw in tightly.

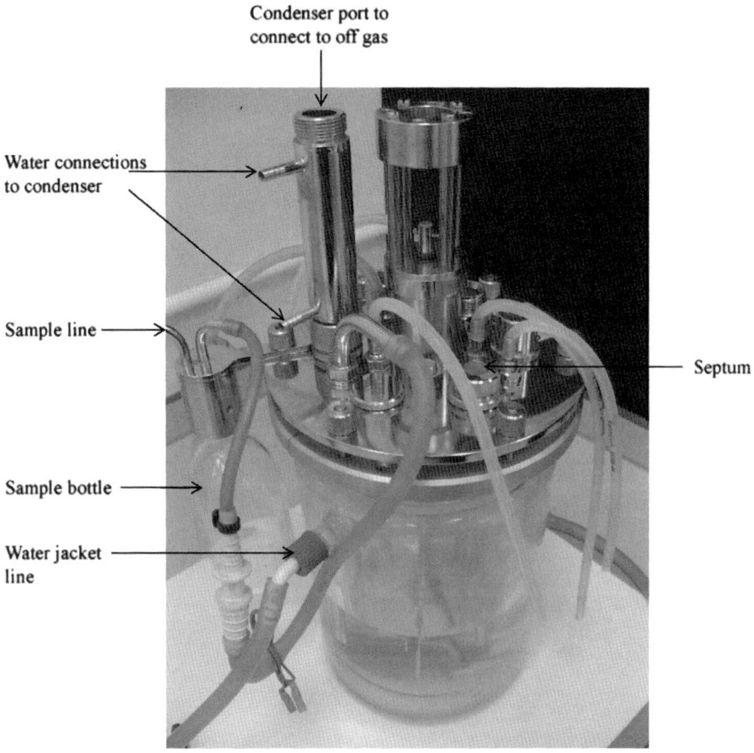

Fig. 2. A photograph showing an assembled Applikon 3 L stirred tank autoclavable bioreactor head plate and vessel in the laboratory of Dr Roslyn Bill, Aston University, UK. The main features are marked.

8. Place the DO probe into the appropriate port after checking the condition of the membrane and that 1 mL of electrolyte is present. Screw the probe in tightly.

9. Place the Optek probe into the appropriate port after checking it is clean and that the O-rings are in good condition. Ensure that the probe is screwed in tightly with the optical path facing towards the outside of the reactor in order to minimize the chance of trapping air bubbles. Cover the top with a protective cap or foil to keep the connections dry.

10. Screw in the condenser to the condenser port after checking the integrity of the O-rings and attach silicon tubing plus a 0.2-μm PTFE gas filter to the top. Clamp the tube and cover the end in foil. The exit gas filter should be left unclamped to allow pressure equalization and avoid vessel damage during autoclaving.

11. Attach a short piece of tubing and a 0.2-μm PTFE gas filter to the inlet gas sparger. This inlet gas filter should be left unclamped to allow pressure equalization and avoid vessel damage during autoclaving. Attaching 0.2-μm PTFE gas filters to the inlet and exit gas lines forms a sterile boundary, thus minimizing contaminations.

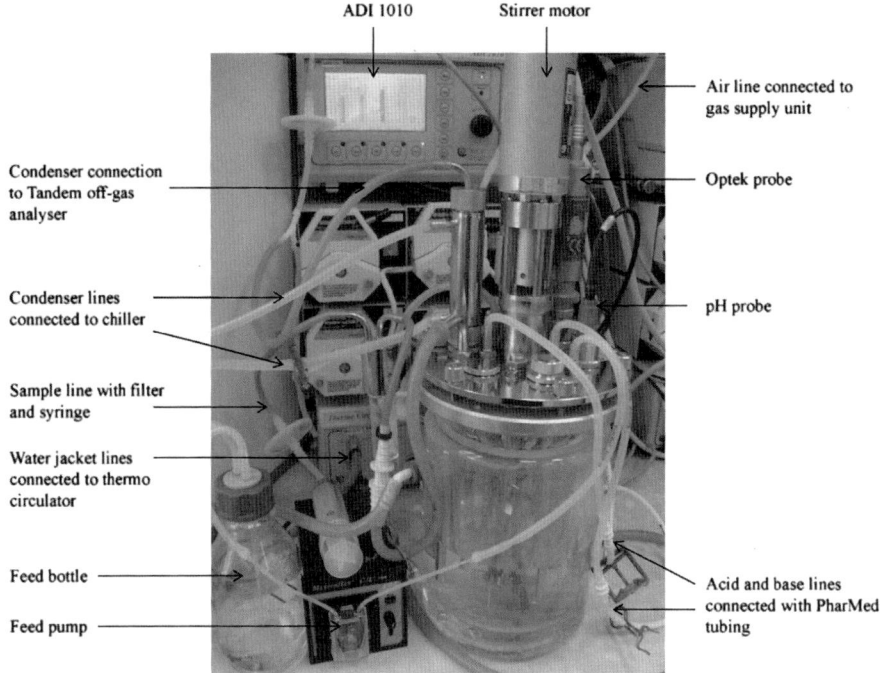

Fig. 3. An annotated photograph showing an Applikon 3 L stirred tank autoclavable bioreactor connected to an ADI 1010 controller, thermo circulator ADI 1018, Tandem off-gas analyser, gas supply unit ADI1026, and Masterflex pumps for acid, base, and feed addition.

12. Attach a 250-mL glass sample bottle to the sample port and place a length of silicon tubing with a filter to the fork. Clamp the tubing between the bioreactor and the sample bottle to prevent transfer of the culture medium during autoclaving.

13. Water jacket lines may be attached to the glass vessel before or after autoclaving. Ensure that they are screwed on tightly to avoid leaks.

14. Place the bioreactor into the autoclave and sterilize at 121°C for 20 min with a slow cool cycle.

3.5. Connecting the Bioreactor to the Control Unit

1. Remove the bioreactor from the autoclave and place next to the control unit (see Fig. 3).

2. Remove the protective cap and connect the DO probe to the ADI 1010 controller and allow to polarize for a minimum of 6 h.

3. Remove the protective cap and connect the pH probe to the ADI 1010 controller.

4. Remove the foil or cap from the top of the Optek probe and connect to the Optek controller.

5. Add 5 mL 50% (w/v) glycerol to the port for the temperature probe (the thermowell tube) and insert the temperature probe.

6. Turn on the chiller and connect the lines to the condenser. Check that the reservoir is filled with deionized water and that it stabilizes to around 6°C.

7. Prepare the Tandem off-gas analyser by connecting a length of silicon tubing. Attach a Y connector to the end of this tubing and attach a smaller length of tubing with a loosened clamp to one of the forks. Tightening this clamp on the open tubing increases the flow of gas to the Tandem-off gas analyser.

8. Attach the forked tubing from the Tandem gas analyser to the exit filter on the condenser gas outlet.

9. Check that the compressor and the 60% oxygen:40% nitrogen cylinder are connected to the ADI 1026 gas supply unit.

10. Attach the gas supply with a length of silicon tubing from the ADI 1026 unit to the filter attached to the sparger on the bioreactor.

11. Attach the stirrer motor to the head plate.

12. Set up acid and base bottles with the pumps connected to the controller equipment. Remove the foil from the two lines on the bioreactor and spray with 70% ethanol before connecting to the lines on the acid and base bottles. Alternatively, make the connection using autoclavable aseptic connectors, such as CPC connections or Rapi-Locks (Applikon).

13. Connect the water jacket lines to the ADI 1018 thermo circulator unit. Water should flow from the bottom connection to the top one.

14. The feed bottles may now be connected to the bioreactor. Insert the tubing into an appropriate peristaltic pump and connect this tubing to the feed line tubing on the bioreactor, spraying the line ends with 70% ethanol. Set the flow rate to the desired value, e.g., 14 mL/h for the 50% (w/v) glycerol feed, and check that the pump is functioning. Reset the pump dose monitors on the controller. Prime the lines and then reclamp to ensure that the culture receives feed as soon as required.

3.6. Setting the Culture Conditions

For a recombinant protein production experiment using *P. pastoris*, commonly used settings are pH 5, 30°C, 30% DO, and a stirrer speed of 700 rpm.

1. Turn on the compressor and set the 60% oxygen:40% nitrogen cylinder to a higher pressure than that of the compressor, e.g. 2.5 bar.

2. Unclamp the air line and on the ADI 1026 gas supply unit, and turn the rotameter until the level is at 1.5 L/min air flow. Check that air bubbles can be seen in the medium escaping from the sparger.

3. Adjust the tightness of the clamp connecting the condenser to the off-gas analyser (see Subheading 3.5, step 7) to ensure that pressure does not build up inside the vessel. Adjust the flow rate of exit gas to the Tandem off-gas analyser to approximately 0.4 L/min by adjusting the clamp on the open end of the forked tubing. If the level is at 1 L/min, the clamp should be loosened to allow more gas to go straight into the laboratory, forcing less through the Tandem.

4. On the ADI 1010, press the stirrer button followed by the Setp button. Using the dial, set the stirrer speed to 700 rpm which is the minimum speed used for the fermentation. Press the Setp button to set the speed. Press the stirrer button again followed by the start button (Start/Stop) to start the stirrer.

5. Check that the medium is uncontaminated once the stirrer starts by observing the bottom of the vessel and checking that the medium is clear and not cloudy.

6. On the ADI 1018 thermo circulator, flip the switch to the "fill" position to replace any jacket water lost during autoclaving.

7. On the ADI 1010 controller, press the temperature button (Temp) and then the Setp button. Using the dial, select the desired temperature, e.g., 30°C. Press the Setp button to set the temperature. Press "Temp" followed by the start (Start/Stop) button.

8. Once the water jacket is full, flip the switch on the thermo circulator ADI 1018 to the "on" position to start heating the water jacket to 30°C.

9. In a laminar flow cabinet, using a 5-mL syringe and needle, withdraw 4.35 mL PTM_1 trace salts. Push the needle through the septum on the head plate of the bioreactor and add the solution. For additional security, pierce the septum through an alcohol swab.

10. Once the temperature has reached 30°C (displayed on the ADI 1010 screen), unclamp the acid and base lines, checking along the lines to ensure that all clamps are removed as this can cause lines to unattach and spray acid or base.

11. To prime the base line, press the pH button followed by the Setp button. Using the dial, select the desired pH, e.g. pH 5. Press the Setp button to set the pH, and then press the pH button followed by the start button (Start/Stop). The base pump starts pumping ammonium hydroxide into the bioreactor as BSM medium has a pH of approximately 1. Monitor until the pH reaches 5. A note of the volume of base and any acid added at this point should be made for future reference.

12. The acid line (and antifoam line if used) should also now be primed, ensuring that the culture receives control reagents as soon as they are demanded by the control loop.

13. Zero the Optek probe to monitor the biomass of the culture. On the Optek fermenter controller, press "clear" and use the arrow buttons to scroll down to maintenance. Press "enter" and scroll down to system zero settings. Press "enter" and select the Optek probe to zero. Press "enter" and then press "enter" again to set the probe to zero. Press "clear" twice and scroll back up to measurement display.

14. Leave the bioreactor running at these settings for an hour to allow it to stabilize.

15. Calibrate the DO probe by pressing the DO button followed by the calibrate button (Calib). Using the dial, enter the DO as 100%. Press "Calib" to set the DO at the current conditions in the bioreactor as 100%. Check that the ADI 1010 display for DO reads 100%.

16. The calibration slope should be between 8 and 15 and the current around 67 nA (13).

17. The 0% setting can be checked either by sparging with nitrogen instead of air until the DO drops to 0% or by quickly disconnecting the DO probe for no longer than a minute and checking that the DO drops to 0% (13). If sparging with nitrogen, reattach the air supply and wait for the settings to return to 100%.

18. To set the DO set point, e.g., to 30%, press the DO button followed by the Setp button. Use the dial to select 30% and then press the Setp button. As each bioprocess differs slightly, DO settings need to be optimized to ensure that the set point is maintained. See the Notes in section 4 for details of how to fine tune the DO control.

3.7. Starting the Run

1. Antifoam may be added at the beginning of the run or when required, as foaming occurs later in the process. The most appropriate type of antifoam and concentration used should be determined as it can influence the $k_L a$ in the bioreactor and the growth of the cells and may affect the yield of recombinant protein produced ((14) and Chapter 9).

2. Measure the optical density of the seed culture at 595 nm using a spectrophotometer.

3. Calculate the volume of seed culture required to inoculate 1 L BSM to the desired OD, e.g., for OD_{595} 0.5, divide 0.5 by the optical density of the seed culture and multiply by the volume in mL, i.e., 1,000.

4. Open BioXpert software and select the user. Select "run" from the top menu bar and then select "new." Enter the details of the run in the comments section.

5. Before saving the file, ensure that the bioreactor settings are correct.

6. Using a syringe and needle, aseptically dispense the required volume of culture through the septum on the head plate of the bioreactor.

7. Immediately save the file in BioXpert. The run has now started and the various parameters can be displayed on the trace by selecting them from the bar below.

3.8. Sampling

1. Attach a 50-mL syringe to the filter on the feed port.

2. Remove the glass sample bottle from the sample port and replace it with a 20-mL sample tube, spraying the inside of the stainless steel lid with 70% ethanol. Unclamp the sample line and using the syringe draw a sample for the $t=0$ h time point into the tube. The culture remaining in the sample line is discarded prior to the next sample being taken.

3. Reclamp the line and replace the sample tube with a new one.

4. If it is not possible to analyze the sample immediately (which is preferred for biomass analysis), store the sample at −20°C until needed.

5. It is recommended that samples are taken at least at $t=0, 4$, and 18 h during the batch phase. Take a sample before beginning the fed-batch phase and also before starvation. Take a sample at the start of the induction phase and each time the flow rate is increased. Continue taking samples every 4–6 h for the remainder of the run and a further sample at the end of the run (15).

3.9. Fed Batch, Starvation, and Induction Phases

1. Once the cells have consumed all the glycerol in the medium, indicated by a DO spike on the BioXpert trace, the fed-batch phase should be started. This is usually observed between $t=20$ and 24 h post inoculation.

2. Start the glycerol feed pump. Ensure that the solution is dripping into the vessel. Allow this feed to continue for approximately 4 h to increase biomass prior to induction.

3. Stop the pump and leave the culture for an hour without any feed to ensure that all glycerol in the medium is utilized.

4. Begin the induction by unclamping the methanol feed line and starting the pump, checking that the solution is dripping into the vessel. The flow rate can be adjusted as required as the run progresses.

5. If preferred, this whole process can be automated with BioXpert version 2. Data from the bioreactor controller, off-gas analyser, optical density system, and many other external analyzers can be incorporated into recipes. Such recipes include process triggers, such as a DO spike, to trigger glucose feed on demand from the DO control loop and a methanol feed after a timed starvation phase. A standard *Pichia* template (Applikon) can be edited as required.

3.10. Cleaning the Bioreactor

1. Once the run is complete and all required samples have been taken, end the run on BioXpert. This causes all temperature, pH, and DO control to cease.

2. On the ADI 1010, turn the stirrer off by pressing "stirrer" followed by Start/Stop button.

3. Turn off the thermo circulator by flipping the switch to the "off" position. Turn off the condenser.

4. Reduce the air flow on the rotameters to 0 and turn off the compressor and gas cylinder.

5. Clamp all lines and disconnect them. Remove the water jacket lines from the thermo circulator. Remove all cables connecting the probes to the controllers and replace protective caps or cover in foil.

6. Replace the glass sample bottle on the feed port.

7. Autoclave the bioreactor.

8. Remove the probes from the head plate and clean using warm water. Ensure that all O-rings are present and undamaged. After drying the probes, check that the membrane of the DO probe is clean and if necessary use a toothbrush. Replace the cap protecting the membrane module and store the probe. Replace the protective cap for the base of the pH probe after ensuring that it is half full with 3 M KCl and store.

9. Unscrew the head plate from the bioreactor and clean with warm water. Rinse out the lines with running warm water.

10. Discard the autoclaved culture by pouring down a sink and clean using warm water to clean the glass vessel.

11. Leave both the head plate and vessel to dry before reassembling and storing.

4. Note

Maintaining DO set point: If the DO drops below the set point during a process, the flow rate of air into the vessel may be increased. It may also be necessary to put the stirrer into a cascade. When the culture demands oxygen, the stirrer speed can then increase to provide further aeration. This in combination with the mass flow controller (MFC), which increases the percentage of air fed from the cylinder when DO is low, can maintain the set point. *PID* settings can also be tuned to improve control, where P is the proportional value and determines how strongly the controller attempts to reach the set point in proportion to how far it is from the set point. If it is too high, the controller will overshoot the set point. This needs to be adjusted so that there is as little overshoot as possible, but the set point is reached within a reasonable time. I is the integral value and determines the time over which P acts and can be tuned after P is set. The larger the I value, the smaller the adjustment will be, as the time for the action to occur is longer. D is the derivative and is not often altered. Examples of stirrer, MFC, and PID settings are given here.

1. *Putting the stirrer into cascade and setting limits*

    (a) Press the Menu button on the ADI 1010 and use the dial scroll to configuration. Press "menu" and then press the Start/Stop button to include the stirrer. Pushing this again will exclude the stirrer.

    (b) To set the stirrer limits once the stirrer is in cascade, press the stirrer button followed by Setp. Set this to the minimum stirrer speed required, e.g., 700 rpm, and press Setp. Press Setp again to return to the main menu.

    (c) Press the Stirrer button and then the Limit button. Using the dial, select 700 rpm for the lower limit—this should be the lowest selectable value. Press the Limit button again and then set the upper limit to 1,250 rpm using the dial.

    (d) Once the run has been started on BioXpert, these settings cannot be altered and the stirrer speed cannot be altered. Press the DO button followed by Start immediately after starting a run and the stirrer will run at 700 rpm until more oxygen is demanded.

2. *Setting the MFC*

    (a) The MFC of the ADI 1026 gas supply unit can output 5 L/min of gas; therefore, in order to obtain, for example, 2 L/min of air in the bioreactors, the MFC settings would need to be at 40%.

    (b) Press the Menu button on the ADI 1010, then select output control using the dial, and press the Menu button. Highlight "O_2 MFC" and select it by pressing "menu."

    (c) Change the upper limit to 40% using the dial and the lower limit to 0%. Press the Menu button to confirm the settings.

3. *PID settings*

    (a) Press the Menu button on the ADI 1010 and scroll to controller settings. Press the Menu button and scroll to DO controller. Select with the Menu button and use the dial to adjust the PID values followed by Menu to confirm them.

    (b) Example settings are $P = 7$, $I = 1,500$, and $D = 0$.

References

1. Macauley-Patrick S, Fazenda ML, McNeil B, Harvey LM (2005) Heterologous protein production using the *Pichia pastoris* expression system. Yeast 22:249–270
2. Chen Y, Cino J, Hart G, Freedman G, White C, Komives E (1996) High protein expression in fermentation of recombinant *Pichia pastoris* by a fed-batch process. Process Biochem 32:107–111
3. Hjersted JL, Henson MA (2006) Optimization of fed-batch *Saccharomyces cerevisiae* fermentation using dynamic flux balance models. Biotechnol Prog 5:1239–1248
4. Holmes WJ, Darby RAJ, Wilks MDB, Smith R, Bill RM (2009) Developing a scalable model of recombinant protein yield from *Pichia pastoris*: the influence of culture conditions, biomass and induction regime. Microb Cell Fact 8:35
5. Zhang W, Hywood Potter K, Plantz B, Schlegel V, Smith L, Meagher M (2003) *Pichia pastoris* fermentation with mixed-feeds of glycerol and methanol: growth kinetics and production improvement. J Ind Microbiol Biotechnol 30:210–215
6. Calik P, Bayraktar E, Inankur B, Soyaslan E, Sahin M, Taspinar H, Acik E, Yilmaz R, Özdamar T (2010) Influence of pH on recombinant human growth hormone production by *Pichia pastoris*. Chem Technol Biotechnol 85:1628–1635
7. Fraser NJ (2006) Expression and functional purification of a glycosylation deficient version of the human adenosine 2a receptor for structural studies. Protein Expr Purif 49:129–137
8. Jamshad M, Rajesh S, Stamataki Z, McKeating JA, Dafforn TR, Overduin M, Bill RM (2008) Structural characterization of recombinant human CD81 produced in *Pichia pastoris*. Protein Expr Purif 57:206–216
9. Dragosits M, Stadlmann J, Albiol J, Baumann K, Maurer M, Gasser B, Sauer M, Altmann F, Ferrer P, Mattanovich D (2009) The effect of temperature on the proteome of recombinant *Pichia pastoris*. J Proteome Res 8:1380–1392
10. Shuler ML, Karagi F (2008) Bioprocess engineering basic concepts, 2 edn. Prentice Hall International Edition
11. Invitrogen: EasySelect *Pichia* expression kit
12. Invitrogen: *Pichia* fermentation process guidelines
13. O'Brien B, Britt K, Audino D (2006) Applikon bioreactor operation SOP
14. Routledge SJ, Hewitt CJ, Bora N, Bill RM (2011) Antifoam addition to shake flask cultures of recombinant *Pichia pastoris* increases yield. Microb Cell Fact 10:17
15. Nyblom M, Öberg F, Petersson-Lindkvist K, Findlay H, Wikström J, Karlsson A, Hansson O, Booth PJ, Bill RM, Neutze R, Hedfalk K (2007) Exceptional overproduction of a functional human membrane protein. Protein Expr Purif 56:110–120

Chapter 11

The Implementation of a Design of Experiments Strategy to Increase Recombinant Protein Yields in Yeast (Review)

Nagamani Bora, Zharain Bawa, Roslyn M. Bill, and Martin D.B. Wilks

Abstract

Biological processes are subject to the influence of numerous factors and their interactions, which may be non-linear in nature. In a recombinant protein production experiment, understanding the relative importance of these factors, and their influence on the yield and quality of the recombinant protein being produced, is an essential part of its optimisation. In many cases, implementing a design of experiments (DoE) approach has delivered this understanding. This chapter aims to provide the reader with useful pointers in applying a DoE strategy to improve the yields of recombinant yeast cultures.

Key words: Design of experiments, Process optimisation, Process development, Process characterisation

1. Implementing a Design of Experiments Approach

The design of experiments (DoE) approach involves the systematic application of statistics to an experimental set-up in order to determine how combinations of a series of input parameters or "factors" set at different "levels" (such as culture temperatures of 20, 25, and 30°C, pH of 5, 6, and 7, and dissolved oxygen concentrations of 30, 40, and 50%) affect an output or "response" (such as recombinant protein yield) (1). DoE is, therefore, an effective way of investigating the impact of multiple conditions while reducing the overall number of experiments, without compromising the quality of the data. Information on the relationship between the factors and the response is extracted in the form of an equation: the use of a statistically robust design means that it is not necessary to perform experiments to examine all possible combinations of factors and levels in order to obtain the equation. In Subheading 2.5, we discuss a recent study exploring three factors set at three levels. The statistical

Roslyn M. Bill (ed.), *Recombinant Protein Production in Yeast: Methods and Protocols*, Methods in Molecular Biology, vol. 866, DOI 10.1007/978-1-61779-770-5_11, © Springer Science+Business Media, LLC 2012

design used required only 13 experimental combinations out of a possible 27 to be examined in order to identify the optimal relationship between the response (in this case, the yield of recombinant green fluorescent protein secreted from the yeast, *Pichia pastoris*) and the factors (temperature, pH, and dissolved oxygen concentration) (2).

In a typical DoE set-up, the factors to be tested, the number of levels, the number of replicates to be performed (e.g. $n = 3$), and the layout of the experiment are specified in a design matrix (see Subheading 2.5 and Note 1). Statistical analysis then fits the response, derived by running the specific experimental combinations defined by the matrix, to a model (which may be linear or nonlinear) and quantitatively determines the effect of each factor on this response. The use of replicates means that the amount of error in the model can be determined as well as whether, or not, any lack of fit present is statistically significant. DoE, therefore, offers many benefits over more traditional experimental approaches of varying one factor at a time (OFAT), which are typically inefficient, expensive, and time consuming (3).

DoE was first proposed as an alternative to OFAT by Sir Ronald A. Fisher in 1935 (4), who based his approach on the statistical method known as "analysis of variance" (ANOVA). It was later used by Genichi Taguchi in the 1950s to improve the quality of manufactured goods and is now widely implemented in modern biotechnological applications (5). DoE as a general strategy is typically involved in both the early and late stages of industrial bioprocess development. More specifically, DoE is seen as being integral to the process of securing regulatory approval for products from organisations, such as the US Food and Drug Administration (see http://www.fda.gov/regulatoryinformation/guidances/ucm128003.htm). In the following sections, the application of DoE to screening, characterisation, and optimisation of protein production experiments is introduced, followed by an overview of appropriate experimental set-ups.

1.1. Screening for Key Factors

Screening designs are used to reduce the factors under initial consideration (which could be 7–12 or more, based on previous experimentation and guidance from the literature (6)) to a shortlist of three to five that warrant further, more detailed study (2). Typically, fractional factorial designs are used at this screening stage, where a "fraction" of the experimental runs are selected from a full factorial design. This allows for a cheap and rapid investigation but may affect the data quality. The compromise between the size of the fraction and the quality of the data can be judged by checking the resolution of the design: a design of at least resolution V is typically chosen (see Chapter 7 of ref. 1 for a further explanation of design resolutions). Implicit in this type of design is that information on how the interactions between factors affect the response is confounded (i.e. distorted). However, data on the main effect of each factor on the response are of sufficient quality to make a judgement about a factor's inclusion or exclusion from subsequent experimentation. Overall, the outcome of a screening exercise should be the identification of the factors that warrant further study, as well as an understanding of their appropriate experimental ranges.

1.2. Process Characterisation

The primary goal of process characterisation is to identify and quantify the influence of the key factors, typically as part of a plan for process improvement. Characterisation confirms the identities of the factors influencing the response of a process (e.g. protein yield, functional activity, or stability) and enables a prediction of the optimal response under a range of operating conditions. The investment of time and resources at this stage results in better process understanding and improved reproducibility and may reduce delays in costly regulatory procedures.

1.3. Process Optimisation

Since a recombinant protein production experiment is a multiphase, multi-component process, protein yield, as well as other responses, such as stability and activity, can be influenced by a wide range of factors including the composition of the culture medium, its pH, the culture temperature, the availability of dissolved oxygen in the medium, and the details of the induction regime (e.g. concentration of inducer as well as the point and duration of induction). In the process optimisation stage, the goal is to "zoom in" on a particular portion of the design space or, by changing the design used, to model any non-linear behaviour observed in the previous stages (e.g. by using the "response surface method"; Subheading 2.4.3). By using DoE, process optimisation becomes more systematic and informative by enabling different levels of each factor and its interactions to be related to the response. In an iterative process, data from one round of DoE results in a model that provides the information for an improved design in subsequent rounds. Table 1 gives some examples of how DoE has been used to improve a range of different bioprocesses, including recombinant protein production experiments.

2. Experimental Set-Up

Devising and analysing a DoE have been considerably simplified in recent years with the advent of a range of specialist software packages, including MiniTab® (http://www.minitab.com), Modde® (http://www.umetrics.com), ECHIP® (http://www.experimentationbydesign.com/index.php), and Design-Expert® (http://www.stateease.com/software.html). These packages are well supported by their providers (see the Web sites above for further information).

Table 1
Examples of DoE in bioprocess improvement

Protein	Goal of DoE	Statistical method used	References
Recombinant erythropoietin (from *P. pastoris* culture)	Maximising protein yield as a function of the temperature, pH, and dissolved oxygen concentration of the culture medium	Response surface method (Box–Behnken)	Bora and Bill (unpublished)
Recombinant green fluorescent protein (from *P. pastoris* culture)	Maximising protein yield as a function of the temperature, pH, and dissolved oxygen concentration of the culture medium	Response surface method (Box–Behnken)	(2)
Polyglutamic acid isolated from *Bacillus subtilis*	Maximising polyglutamic acid yield as a function of the composition of the growth medium	Fractional factorial design and response surface method	(16)
Recombinant Fab' fragment (from *Escherichia coli* culture)	Maximising yield as a function of agitation rate and dissolved oxygen concentration	Full factorial (2^2) design	(17)
Clavulanic acid from *Streptomyces clavuligerus*	Maximising clavulanic acid yield by optimising the composition of the growth medium	Screening by fractional factorial design and optimisation by response surface method	(18)
Recombinant cystatin C mutant (from *P. pastoris* cultures)	Maximising yield and protein glycosylation as a function of three nitrogen sources	Full factorial (2^3) design	(19)
Neomycin isolated from *Streptomyces marinensis*	Maximising neomycin yield by optimising the composition of the growth medium	Full factorial design and response surface method	(20)

Before starting a DoE, the experimental goals and criteria for success should be clearly articulated. A relevant example would be the goal of determining the key factors influencing a response, such as recombinant protein yield, and then to use that information to maximise the yield, as measured in mg/L (Table 1). Only once these goals and criteria are defined can a valid DoE strategy be developed, including a plan of action in the event when the experiments do not turn out as expected. The effect of selected factors on a process response is then examined at a number of levels, depending on the experimental design chosen. It should be noted that the temptation to add a large number of factors or responses just to see how they change should be tempered by the fact that this may divert focus from those that are critical to meeting the goals of the DoE, and therefore should be avoided. In the following sections, the key components in setting up a DoE are considered.

2.1. Factor Selection

Factors are usually variables, which can have defined set points. They might include pH, temperature, dissolved oxygen concentration, or the concentration of medium components. Input factors may also be an "attribute", e.g. the presence or absence of a medium component at a level that does not vary. Other factors, which may or may not be controllable and which are referred to as "noise factors", should be considered in the DoE (see Note 2). The presence of noise during the experiment can distort the results to the extent that incorrect conclusions are drawn. Their effect may, therefore, be minimised by using "blocking" or "randomisation" in the design (see Chapter 2 of ref. 1 for further details). "Blocking" mitigates "categorical" noise, e.g. that introduced by using "bioreactor 2" in some experimental runs instead of "bioreactor 1". "Randomisation" mitigates "variable" noise, e.g. day-to-day variations in laboratory temperature.

2.2. Level Selection

For the simplest designs, known as 2^k designs, each of k factors is examined at two different levels, coded as −1 (for the low level) and +1 (for the high level) in a design matrix. This type of design can also be modified to accommodate many more levels. However, it is important to bear in mind that examining certain levels may not be biologically practical. For example, a growth medium with a very low pH may inhibit the growth of the organism being studied while maintaining very high dissolved oxygen concentrations may not be experimentally feasible. Since the difference in response observed experimentally is related to the difference in the levels of each factor, an equation can be derived that describes the relative importance of each factor on any change to that response.

2.3. Response Selection

It is possible to carry out DoE, where the response is an attribute (7) (e.g. the protein produced is functional or not), but most commonly the response can be measured on a continuous variable scale. Protein yield, protein activity, and culture density fall into this category.

2.4. Experimental Design Selection

The choice of experimental design, as discussed in the next section, is dependent on the purpose of the DoE (screening, characterisation, or optimisation) and the number of factors under consideration, as summarised in Table 2.

Table 2
An overview of statistical designs and when to use them

Number of factors	Screening	Characterisation	Optimisation
1	Not applicable for a single factor	Linear regression or, in cases where there is no linear fit, non-linear regression	Linear or non-linear regression
2–4	Full factorial	Full factorial	Full factorial (for linear response) or response surface method (for non-linear response)
5 or more	Fractional factorial	Full factorial on selected factors (usually, <4)	Full factorial (for linear response) or response surface method (for non-linear response)

2.4.1. Factorial Designs

Full factorial designs (e.g. 2^k designs) can be used for screening a small number of factors (≤4) in order to identify the most significant ones, but can also be used sequentially to model and refine a process. Each factor can have two or more levels and the design generated will include all possible combinations of the factors and levels. In contrast, fractional factorials are more efficient designs used to screen a large number of factors (≥5) to find the few that are significant, but compromise on the quality of the information on the interactions between the factors. Consequently, full factorials should be used to estimate the effects of interactions, which may be missed in a fractional design.

In cases where a large number of factors is to be studied while minimising the experimental runs, Plackett–Burman designs (8) may be considered. Alternatively, a d-optimal approach may be suitable, as it allows a subset of experimental runs to be selected (9). The d-optimal design also allows the inclusion of both quantitative and qualitative (attribute) factors with a mixed number of levels. Analysis of factorial designs is typically done using ANOVA (1), which leads to a first-degree polynomial equation describing the factors that influence the response of interest. However, full factorial designs may also be analysed using regression (see Subheading 2.4.3).

2.4.2. Taguchi Designs

Taguchi's orthogonal arrays (1), which were originally created before the widespread use of DoE software, are highly fractional designs that can be used to estimate the main effects using only a few experimental runs. These designs are not only applicable to two-level factorial experiments, but also can investigate the main effect of a factor with more than two levels. Designs are also available to investigate the effects when the factors do not have the same number of levels. As with Plackett–Burman designs, these designs require the experimenter to compromise on data describing any interaction effects. Taguchi designs are often focused on reducing the sensitivity of a response to noise. A recent example of the use of this type of approach is in the improvement of biological assays (10).

2.4.3. Response Surface Method Designs

Factorial designs are sufficient to determine which factors have an impact on the response of interest. Once these have been identified, a more complex design can be implemented to generate a second-degree polynomial equation, which can be used to maximise, minimise, or achieve a specific response. Regression models are used for analysis of the response, as quantifying the relationship between the response and the factors is of the most interest, rather than the identification of the important factors: this is known as the response surface method (RSM) (11). Once the resultant equation has been validated, the behaviour of a process can be predicted, for example in maximising protein yield (Table 1).

In order to analyse response surfaces, special experimental designs are used that help the experimenter fit the second-order equation to the response in the minimum number of runs. Examples of these designs include the Central Composite design (CCD) and the Box–Behnken design (BBD) (11). CCD is a two-level, full or fractional factorial design augmented with a number of centre points and other chosen runs (12). BBD is similar in concept to Plackett–Burman designs, but with factors at three levels. Note that a full factorial design, with all factors at three levels, would also provide all required regression parameters. However, this type of design is expensive to use, requiring 27 runs compared, for example, with the 13 required in a BBD and 15 in a CCD (13). One further advantage of BBD in biological applications is that it does not contain factor combinations for which all factors are simultaneously at their highest or lowest levels, thus avoiding experiments that need to be performed under extreme conditions. However, if an experimenter is interested in the responses at the extremes, BBD may not be suitable.

2.5. Data Analysis: A Case Study

A BBD was used to optimise the yield of recombinant green fluorescent protein (secreted from *P. pastoris*) as a function of the three most commonly varied process parameters: culture temperature (T), pH, and percentage of dissolved oxygen in the culture medium (DO) (2). Based on the results of a first optimisation run (see http://www.ncbi.nlm.nih.gov/pmc/articles/PMC2717918/#supplementary-material-sec), the three factors (T, pH, DO) were each varied at three levels, coded as –1 (lowest value), 0 (middle value), and +1 (highest value); MiniTab® statistical software (version 15.1.1.0) was used to construct the experimental matrix shown in Table 3.

Table 3
Factors and measurable responses for the model building experiments

Inputs (controlled online)			Measurable responses (measured offline)				
T (°C)	pH	DO (%)	OD_{595}	RFU (mL^{-1})	Specific RFU (mL^{-1} $OD_{595}$$^{-1}$)	Specific yield (ng/mL OD_{595})	SD; $n=3$ (ng/mL OD_{595})
19	6	60	20.3	8,651	426.2	127.9	3.2
19	8	60	0.8	1,015	1,268.8	380.6	3.9
19	7	30	13.1	10,984	838.5	251.6	1.3
19	7	90	12.4	9,259	746.7	224.0	2.1
24	6	30	24.4	8,061	330.4	99.1	1.6
24	6	90	16.2	11,951	737.7	221.3	5.6
24	8	30	4.7	1,564	332.8	99.8	1.1
24	8	90	1.3	1,954	1,503.1	450.9	1.3
24	7	60	17.6	21,382	1,214.9	364.5	10.1
29	7	30	24.8	25,392	1,023.9	307.2	0.2
29	8	60	4.4	1,413	321.1	96.3	1.5
29	6	60	21.7	10,349	476.9	143.1	0.3
29	7	90	15.1	17,495	1,158.6	347.6	3.5

The input factors were temperature (T), pH, and % dissolved oxygen (DO). Relative fluorescent units (RFUs) and the optical density at 595 nm (OD_{595}) were measured in triplicate 48 h post induction. The mean values are reported for 1 mL of culture. The standard deviation (SD; $n=3$) is given for the specific yield of the culture, where the conversion factor from RFU to ng was determined by generating a standard curve (adapted from ref. 2)

2.5.1. Model Building

The predictive model generated from the outputs of the matrix is described by Eq. 1 and Fig. 1.

$$\text{Yield } (\text{ng/mL } OD_{595}) = (-21,814.9 + (328.6 \times T)$$
$$+ (5,502.1 \times \text{pH}) - (37.8 \times \text{DO})$$
$$- (325.6 \times \text{pH}^2) - (47.9 \times T \times \text{pH})$$
$$+ (6.4 \times \text{pH} \times \text{DO})) \times \gamma, \qquad (1)$$

where T = temperature (°C), DO = dissolved oxygen (%), and $\gamma = 0.3$ and is the conversion factor from RFU to ng of protein.

This model was derived in Minitab® (see http://www.minitab.com for a detailed description of its use) by removing insignificant terms from the full model based on their p-values (14). The adjusted R^2 value (R_{adj}^2) for the regression changed as each term was removed, R_{adj}^2 being a modification of R^2 that adjusts for the number of

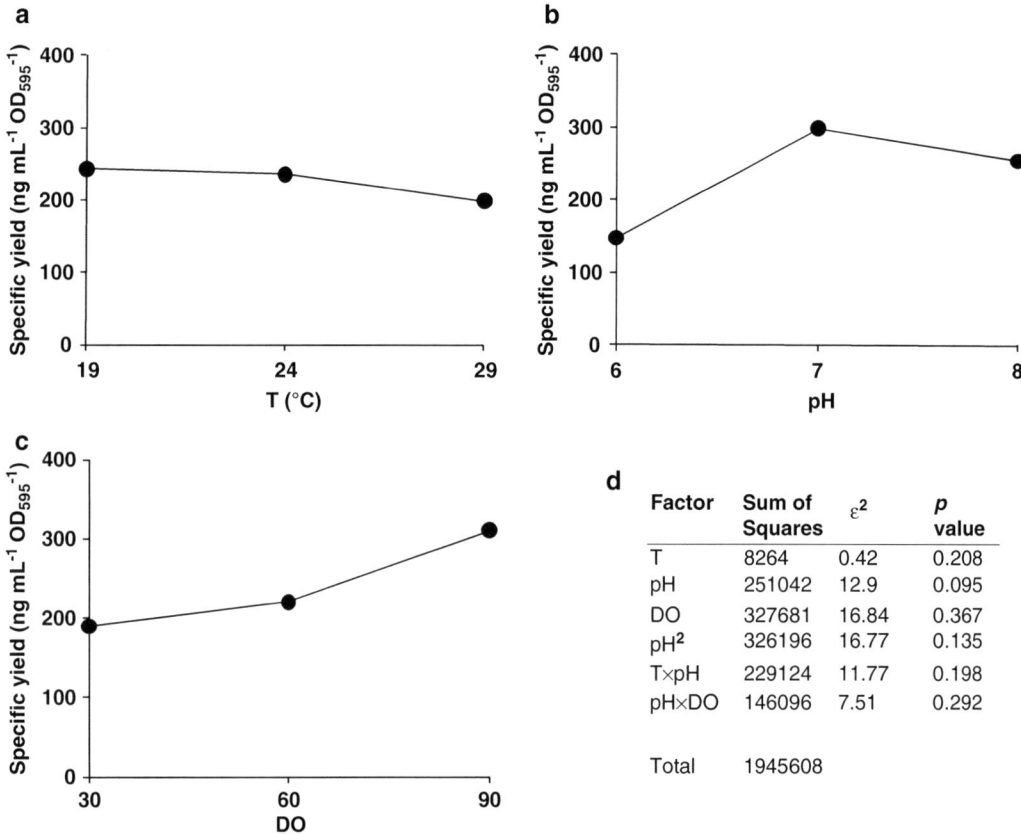

Fig. 1. A main effects plot showing the influence of each of the factors (**a**) T, (**b**) pH, and (**c**) DO on the response (specific yield). Panel (**d**) shows the ε^2 analysis which indicates the influence of each of the factors and their interactions on the model. The value reported for ε^2 is the quotient of the sum of squares for the factor and the total sum of squares (from Table 4) expressed as a percentage.

Factor	Sum of Squares	ε^2	p value
T	8264	0.42	0.208
pH	251042	12.9	0.095
DO	327681	16.84	0.367
pH²	326196	16.77	0.135
T×pH	229124	11.77	0.198
pH×DO	146096	7.51	0.292
Total	1945608		

terms in the model (14). R_{adj}^2 values of 0.160 (full model), 0.115 (one term removed), 0.274 (two terms removed), 0.324 (three terms removed), and 0.292 (four terms removed) indicated that the model with three terms removed was statistically soundest. In Eq. 1, the yield was converted to ng/mL OD$_{595}$ from RFU/mL OD$_{595}$ using an experimentally derived factor.

Yields improved at lower T and higher pH, although at the temperatures tested T did not have a large effect on yield (Fig. 1a), which was highest around pH 7 (Fig. 1b). Yields also increased with increasing DO (Fig. 1c). Figure 1d shows the ε^2 results, which indicate the influence of each of the factors and their interactions within Eq. 1. The data support the view that pH is a key factor as the ε^2 values for pH and pH^2 and the interactions of pH with both T and DO are substantial. DO alone is also important while in contrast the effect of T alone makes a relatively small contribution, in agreement with the main effects plots (Fig. 1).

2.5.2. Model Validation

The results of the statistical validation of this model by ANOVA are shown in Table 4. A recent report suggests that this type of analysis is often missing in published models and that good models from the literature have R^2 values > 0.75 with values below 0.25 being considered poor (15). This suggested that the model was of acceptable quality in line with recent DoE studies of protein production in *E. coli* (15).

The model was also validated experimentally by running the factor combinations shown in Table 5, which had not been used in the model building process, and comparing the fit of the experimental output to the predicted response from the model (Fig. 2).

Nine of the twelve data points were within 40 ng/mL OD_{595} (i.e. within 5–15%) of the predicted value. The three data points outside this range (with T, pH, and DO values of 20, 7.5, 60; 28, 7.5, 90, and 27.5, 6.7, 80) were within 16–25% of the predicted value, and were not correlated in any obvious manner. The experimental conditions leading to the maximum yield were predicted to be 21.5°C, pH 7.6, and DO 90% (Fig. 3), which were confirmed experimentally (2).

Table 4
Statistical significance of the predictive model by ANOVA

Source	Degrees of freedom	Sum of squares	Mean square	F statistic	p value
Regression	6	1,288,405	214,734	1.96	0.217
Linear	3	586,988	223,083	2.04	0.21
Square	1	326,196	326,196	2.98	0.135
Interaction	2	375,221	187,610	1.71	0.258
Residual	6	657,203	109,534		
Total	12	1,945,608			

The statistical significance of the relationship between the predictors and the response of the model was assessed using ANOVA, which employs Fisher's F-test. The goodness of fit of the model is 66%, as determined by the quotient of residual sum of squares/total sum of squares ($R^2 = 0.66$) (adapted from ref. 2)

Table 5
Specification of the input factors for the model validation experiments

T (°C)	pH	DO (%)
20	7.5	60
20	7.7	80
27	8	50
28	7.5	90
28	6	80
23.6	7.25	60
27.5	6.7	80
27.5	6.5	60
27.5	6.3	60
21.5	7.6	20
21.5	7.6	40
21.5	7.6	60

The input factors were temperature (T), pH, and dissolved oxygen (DO) (adapted from ref. 2)

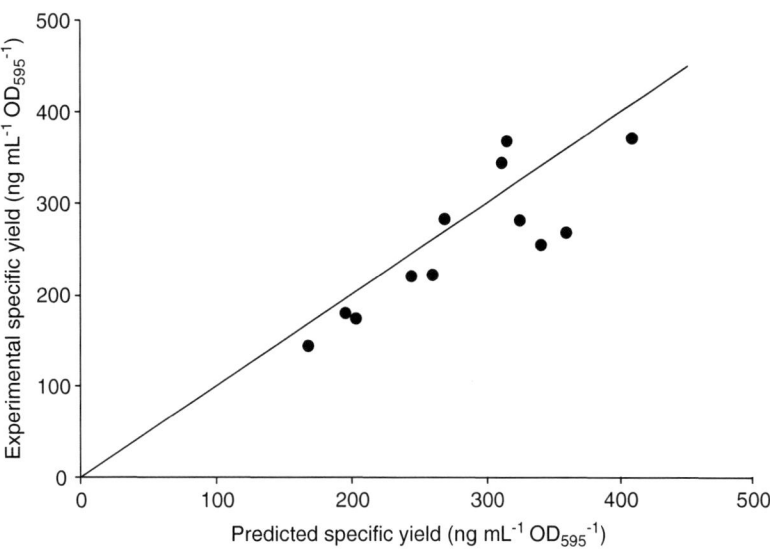

Fig. 2. Demonstration of the predictive capacity of the model. A scatter plot of the predicted versus experimental response is shown. Each check point condition was from within the model design space, but had not been used to build the model. The fit to the line of parity ($y = x$) is shown with $R^2 = 0.57$.

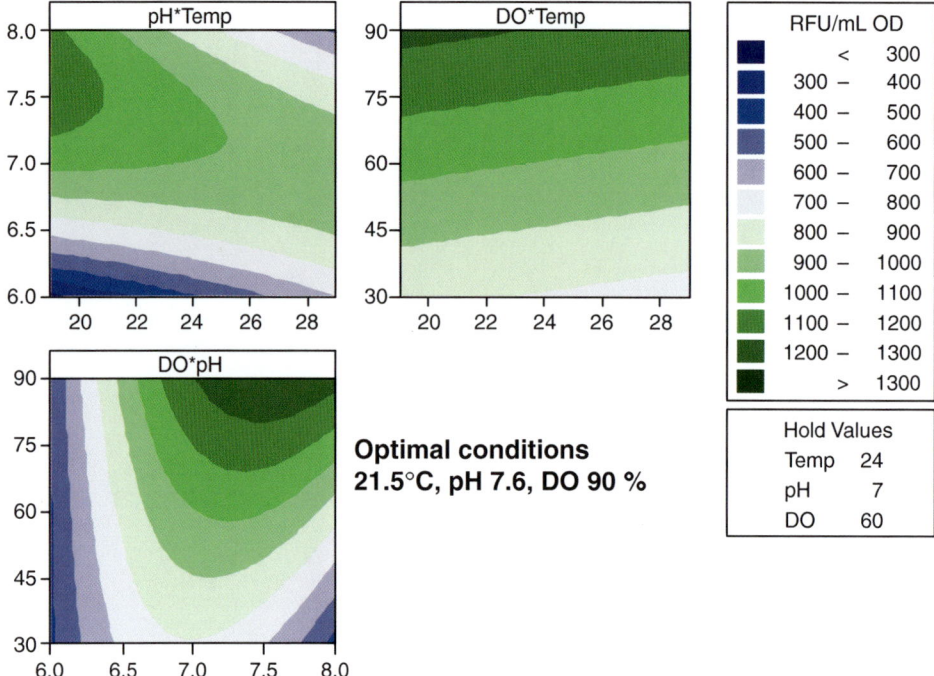

Fig. 3. A response surface contour plot showing how yield per cell changes with each of the input factors. T = temperature (°C), pH = pH, and DO = dissolved oxygen (%). All hold values are the "0" mid-point values in the DoE matrix.

3. Notes

1. Before starting any experimentation, ensure the reliability of any gauges or measurement devices to be used and record any process drifts or changes (such as a change of operator) during the experiment. A minimum of three replicates should be done per experiment. Where possible, use the same starting materials for all experiments. Document all raw output data as well as the averaged data.

2. Noise factors may be categorical (such as noise associated with a change in operator or item of equipment) or random (the ambient temperature or humidity).

References

1. Anthony J (2003) Design of experiments for engineers and scientists. Butterworth-Heinemann, Oxford
2. Holmes WJ, Darby RAJ, Wilks MDB, Smith R, Bill RM (2009) Developing a scalable model of recombinant protein yield from *Pichia pastoris*: the influence of culture conditions, biomass and induction regime. Microb Cell Fact 8:35
3. Czitrom V (1999) One-factor-at-a-time versus designed experiments. Amer Statistician 53:2
4. Fisher RA (1971) The design of experiments, 9th edn. Macmillan, London
5. Rao RS, Kumar CG, Prakasham RS, Hobbs PJ (2008) The Taguchi methodology as a statistical tool for biotechnological applications: a critical appraisal. Biotechnol J 3:510–523

6. Knospel F, Schindler RK, Lubberstedt M, Petzolt S, Gerlach JC, Zeilinger K (2010) Optimization of a serum-free culture medium for mouse embryonic stem cells using design of experiments (DoE) methodology. Cytotechnology 62:557–571

7. Bisgaard S, Fuller HT (1994–1995) Analysis of factorial experiments with defects or defectives as a response. Quality Eng 7:429–443

8. Yuan L-L, Li Y-Q, Wang Y, Zhang X-H, Xu Y-Q (2008) Optimization of critical medium components using response surface methodology for phenazine-1-carboxylic acid production by *Pseudomonas* sp. M-18Q. J Biosci Bioeng 105:232–237

9. de Aguiar PF, Bourguignon B, Khots MS, Massart DL, Phan-Than-Luu R (1995) D-optimal designs. Chemometrics Internat Lab Sys 30:199–210

10. Luo W, Pla-Roca M, Juncker D (2011) Taguchi design-based optimization of sandwich immunoassay microarrays for detecting breast cancer biomarkers. Anal Chem 83:5767–5774

11. Myers RH, Montgomery DC (1995) Response surface methodology: process and product optimization using designed experiments, 1st edn. Wiley, New York

12. Einsfeldt K, Severo Junior JB, Correa Argondizzo AP, Medeiros MA, Alves TL, Almeida RV, Larentis AL (2011) Cloning and expression of protease ClpP from *Streptococcus pneumoniae* in *Escherichia coli*: Study of the influence of kanamycin and IPTG concentration on cell growth, recombinant protein production and plasmid stability. Vaccine. doi:10.1016/j.vaccine.2011.05.073

13. Ferreira SL, Bruns RE, Ferreira HS, Matos GD, David JM, Brandao GC, da Silva EG, Portugal LA, dos Reis PS, Souza AS, dos Santos WN (2007) Box-Behnken design: an alternative for the optimization of analytical methods. Anal Chim Acta 597:179–186

14. Montgomery DC, Peck EA (1982) Introduction to linear regression analysis. Wiley, New York

15. Mandenius CF, Brundin A (2008) Bioprocess optimization using design-of-experiments methodology. Biotechnol Prog 24:1191–1203

16. Shi F, Xu Z, Cen P (2006) Efficient production of poly-gamma-glutamic acid by *Bacillus subtilis* ZJU-7. Appl Biochem Biotechnol 133: 271–282

17. Garcia-Arrazola R, Dawson P, Buchanan I, Doyle B, Fearn T, Titchener-Hooker N, Baganz F (2005) Evaluation of the effects and interactions of mixing and oxygen transfer on the production of Fab' antibody fragments in *Escherichia coli* fermentation with gas blending. Bioprocess Biosyst Eng 27:365–374

18. Wang Y-H, Yang B, Ren J, Dong M-L, Liang D, Xu AL (2005) Optimization of medium composition for the production of clavulanic acid by *Streptomyces clavuligerus*. Process Biochem 40:1161–1166

19. Pritchett J, Baldwin SA (2004) The effect of nitrogen source on yield and glycosylation of a human cystatin C mutant expressed in *Pichia pastoris*. J Ind Microbiol Biotechnol 31: 553–558

20. Adinarayana K, Ellaiah P, Srinivasulu B, Bhavani Devi R, Adinarayana G (2003) Response surface methodological approach to optimize the nutritional parameters for neomycin production by *Streptomyces marinensis* under solid-state fermentation. Process Biochem 38:1565–1572

Chapter 12

Online Analysis and Process Control in Recombinant Protein Production (Review)

Shane M. Palmer and Edmund R.S. Kunji

Abstract

Online analysis and control is essential for efficient and reproducible bioprocesses. A key factor in real-time control is the ability to measure critical variables rapidly. Online in situ measurements are the preferred option and minimize the potential loss of sterility. The challenge is to provide sensors with a good lifespan that withstand harsh bioprocess conditions, remain stable for the duration of a process without the need for recalibration, and offer a suitable working range. In recent decades, many new techniques that promise to extend the possibilities of analysis and control, not only by providing new parameters for analysis, but also through the improvement of accepted, well practiced, measurements have arisen.

Key words: Bioprocess, Monitoring, Analysis, Control, Sensor, Online, In situ

1. Introduction

Every biological process has critical variables that must be measured and controlled. This is true during the developmental stage of a process, where as many parameters as possible need to be measured in order to gain understanding, as well as in established processes in bioreactors, where optimal conditions need to be maintained. Fast and efficient measurements are always preferred, since they allow a more immediate response. Accordingly, the ability to measure parameters online and in situ is of great importance, especially when they are used in closed control loops.

The incentives for improved analysis and control are increased efficiency, productivity, and reproducibility. For effective analysis and control, we require a range of sensors that can measure the state of important process variables. In turn, these sensors must be coupled to a suitably responsive controller to affect an appropriate and stable response.

Roslyn M. Bill (ed.), *Recombinant Protein Production in Yeast: Methods and Protocols*, Methods in Molecular Biology, vol. 866, DOI 10.1007/978-1-61779-770-5_12, © Springer Science+Business Media, LLC 2012

Sensor development stimulates vast amounts of research, but relatively little of this effort is seen to result in robust techniques that are commercially viable or applicable to broad application. For sensors to be reliable in bioprocesses, they have to operate in harsh environments for extended periods of time without the need for recalibration, and with limited sensitivity to changes in process conditions. The need for sensors to be sterilizable, preventing contamination through permeation of components into or out of the sensor provides added difficulty.

Measurement and control of pH, temperature, and dissolved oxygen has become conventional in bioreactors but few other measurements could be considered commonplace with the possible exception of biomass monitoring. At the very least, it would seem desirable to monitor and control substrate levels and have the means to measure specific products more readily.

Although few sensors can be considered mainstream, many other parameters are now routinely measured and controlled using a wide range of technologies. Some of the traditional methods of measuring conventional parameters are also being improved. This chapter aims not only to outline the current status of mature, commonplace techniques but also to review promising areas of research with the potential to challenge accepted techniques in the future.

2. Temperature

One of the most important parameters to control in any process is temperature. The Arrhenius-type relationship between temperature and reaction rate only holds over narrow temperature ranges in most bioprocess applications because biological specimen denature at higher temperatures, stressing further the importance of temperature monitoring. Biological reactions often generate heat and mechanical processes, such as agitation, also contribute; the larger the culture, the greater the contribution. Furthermore, oxygen solubility is dependent on temperature, and temperature readings are required for the correction of other measurements, such as pH.

Almost all modern temperature sensors used in bioprocesses are resistance temperature detectors or thermistors. The most common materials used for these detectors are high purity platinum, nickel, or copper. While thermocouples are still common for high temperature applications, resistance temperature detectors are more accurate and stable at ambient and moderate temperatures (1).

Resistance temperature detectors use the principle that materials exhibit a specific change in electrical resistance as a consequence of a temperature change. In contrast to thermocouples, resistance temperature detectors require a power source to apply a voltage to measure the resistance. A linear resistance–temperature relationship

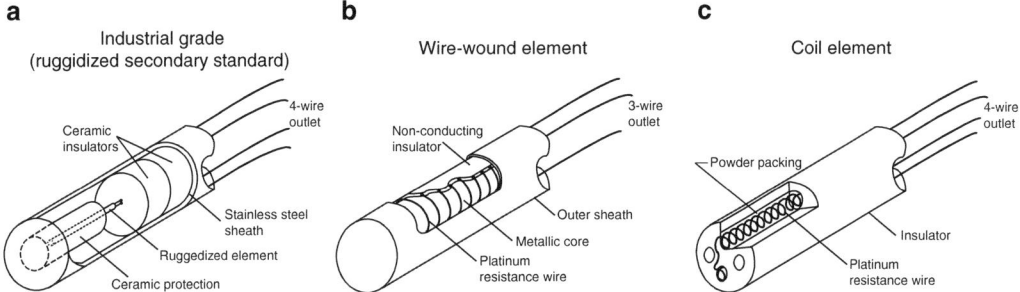

Fig. 1. Resistance temperature detectors. (**a**) Industrial grade elements are used for reference work and offer a good working range. (**b**) Wire-wound and (**c**) coil elements offer improved robustness because the platinum wire can expand without breakage. They also have a lower material cost.

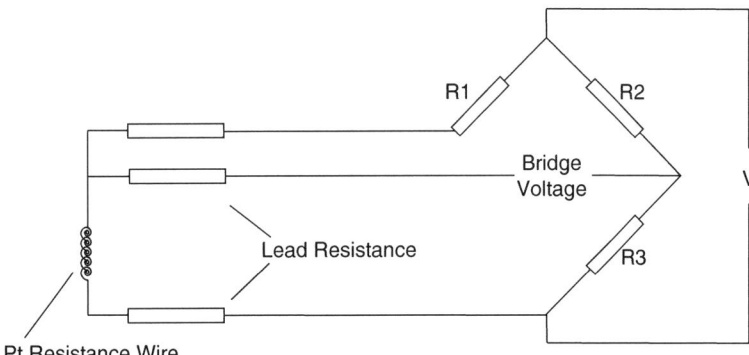

Fig. 2. A three-wire RTD configuration minimizes the lead resistance in a temperature sensor. The two leads from the sensor are placed on adjoining arms, and the assumption is made that the two lead resistances are the same, so high quality connection cables should be used with this configuration. R1, R2, and R3 are known resistances in the Wheatstone bridge circuit. The applied voltage, V, yields a bridge voltage that is proportional to the sensor resistance.

is desirable and platinum has become the industry standard on that basis, having a good linear profile from −184.44°C to +648.88°C (2). Platinum provides the added advantage of being chemically inert. The most common device is a Pt-100 probe, a platinum (Pt) probe that has a resistance of 100 Ω at 0°C (BS EN 60751:1996, taken from IEC 60751:1995). The modern configuration is a wound platinum wire supported in a ceramic or glass insulator. This resistance thermometer unit and the lead connections are further covered by a protective sheath, usually constructed from stainless steel for bioprocess applications.

Several progressions of the Pt-100 sensor can be found in existence, the main advances being the way in which the platinum wires are incorporated in the sensor (Fig. 1) and the way the lead resistance is compensated (Figs. 2 and 3). Element configuration is a choice between ruggedness, range, and cost. Wire-wound (Fig. 1b) and coiled elements (Fig. 1c) are cheaper and are more

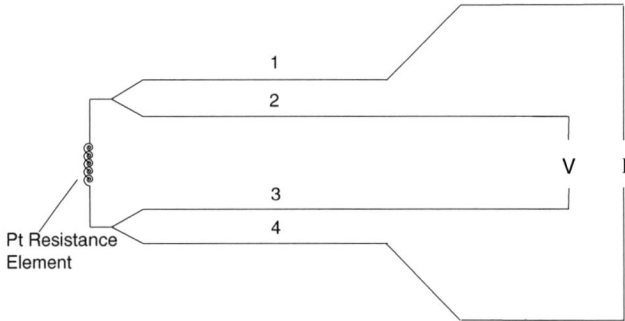

Fig. 3. A four-wire Kelvin RTD configuration used in a temperature sensor. Separate pairs of current carrying (1 and 4) and voltage sensing (2 and 3) electrodes are created. Virtually no current (*I*) flows in the sensing wires and thus the voltage drop in accordance with Ohm's law is very low.

rugged, as they allow for expansion of the coil with temperature change. Industrial grade sensors (Fig. 1a) are more fragile, but have a wider range, typically –200°C to +1,000°C. In any case, the insulator provides protection from vibration and, crucially, protects the platinum from impurities.

The unknown resistance from Pt-100 sensors can be measured using a Wheatstone bridge circuit. The unknown "sensor resistance" is a component in one leg of the bridge circuit, balanced by a second leg of known resistance. To improve accuracy, additional wires are used to create an additional loop that cancels the resistance of the sensor connection lead. By separating the current and voltage electrodes in a Kelvin bridge arrangement, the resistance of the sensor wiring is eliminated. 3-wire connections are still common (Fig. 2), but 4-wire connections are used for very precise applications (Fig. 3). A study of the accuracy and reproducibility of calibration for nineteen 100 Ω resistance temperature detectors was carried out by Wang et al. (1).

3. pH

Most yeast strains will produce acidic and/or basic by-products during metabolism. Consequently, optimal cell growth cannot be achieved without reliable pH control. The most common type of pH sensor is the sterilizable glass electrode. Although they remain the recognized standard, poor mechanical performance of these glass probes is driving research into more robust alternatives. One extensively researched option is the use of optical sensors based on absorbance or fluorescence of pH-sensitive dyes. Despite this effort, overall range remains an issue with optical pH measurement and

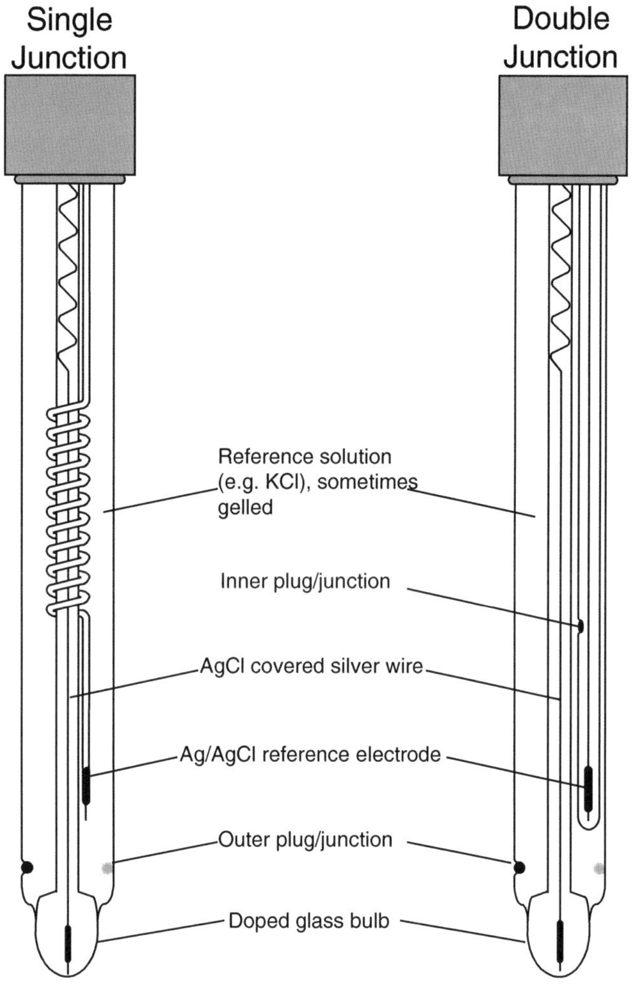

Single
Junction

Double
Junction

Reference solution
(e.g. KCl), sometimes
gelled

Inner plug/junction

AgCl covered silver wire

Ag/AgCl reference electrode

Outer plug/junction

Doped glass bulb

Fig. 4. Single- and double-junction reference electrodes. The additional transfer step of the double-junction probe slows the loss of electrolyte, and reduces maintenance.

the lack of a single sensor for wide application being a concern for bioprocess applications. A notable research focus is novel polymer-based optical sensors as a solid-state option. Ion-selective field-effect transistor (ISFET) sensors are commercially available and are promising, but they have not been extensively tested in bioprocesses.

The traditional glass electrode remains the state-of-the-art sterilizable sensor for pH measurements, even though it has been around for more than 100 years. The most common configuration is the combination probe that incorporates a reference electrode into the same unit as the measuring electrode (Fig. 4). The combination probe is a concentric arrangement of sealed tubes that each contain an electrode in a chamber of electrolyte. This type of pH sensor can be considered analogous to a galvanic cell. The inner measurement tube is separated from the measured solution by a lithium-doped glass bulb designed to screen out many of the ions found in solution with the exception of protons, which are able to migrate through the selective barrier, producing a measurable voltage that is proportional to the pH value of the external solution. Reference measurements are taken in the outside tube where the reference electrode sits in a neutral electrolyte solution, usually potassium chloride. The outer reference chamber is not closed because a porous membrane or ceramic plug, known as a frit, is placed in the outer wall to allow exchange of ions with the measured solution. As such, the outer electrode provides a stable zero-voltage that completes the circuit. The frit, which allows ion exchange, is analogous to the salt bridge in a galvanic cell.

In addition to fragility, the loss of electrolyte from the reference cell is a defining factor in the longevity of the glass pH sensor. During use K^+ and Cl^- ions are slowly exchanged and released without affecting most bioprocesses, but the electrolyte becomes contaminated and must be replaced. With off-line sensors the electrolyte is often replaced through filling holes, but with online sensors alternative steps are taken to limit the need for such maintenance. One solution is to gel the electrolyte, but this only slows leakage and does not affect the rate of ion transfer. Since it is difficult to refill the gel, overall sensor lifespan is shortened. Although there is an additional cost, some online sensors have double junctions where an additional sheath with a further porous plug is placed around the reference electrode. Single- and double-junction sensors are illustrated in Fig. 4. Favourable attributes of glass electrodes such as these include the fact that their responses can be described mathematically using a form of the Nernst equation Eq. 1, where E is the galvanic cell potential and E_0 the standard galvanic cell potential at the temperature of interest. kT in this form of the equation is the Nernst factor Eq. 2 and represents the change in total potential for every unit change of pH. If the absolute temperature (T) is constant, kT is constant. If T changes, then the Nernst factor, also termed the slope factor, changes and shows how the potential output of the pH electrode is dependent on temperature. Glass electrodes also have immunity to redox interference.

$$E = E_0 - kT \cdot \text{pH} \tag{1}$$

$$kT = 2.3 \frac{RT}{F} \tag{2}$$

Where R represents the universal gas constant and F is the Faraday constant.

3.1. Metal/Metal Oxide pH Electrodes

Unlike glass electrodes where there is a membrane potential, metal/metal oxide (MMO) electrodes generate a potential from the resultant equilibrium between a sparingly soluble salt and its saturated solution (3). MMO electrodes have been constructed from a wide range of materials, including Pt/PtO_2, W/W_2O_3, Pb/PbO_2, Ru/RuO_2, Ir/IrO_2, and Sb/SbO_2 (4). Many methods have been used to deposit the oxide layer onto the metal electrodes, including electrochemical growth (5), electrodeposition (6, 7), reactive sputtering (8), and thermal preparation (9). Of particular note, is the solid-state Ir/IrO_2 electrode where the iridium oxide film is prepared by carbonate-melt oxidation (10). This method of preparation overcomes the problems of signal drift reported with other techniques. This sensor has good stability over a 2.5-year test period with a working range of pH 1–13, and offers a fast Nernstian response of 90% in under 1 s (11). The sensor is reported to operate at temperatures of up to 200°C in aqueous solution, and at high pressure (8). The Sensirox INpHO electrode is an existing commercial example of an MMO electrode.

3.2. Optical pH Sensors

Optical fibre pH sensors are also commercially available and Ocean Optics produces a reflective film sensor that can be used for turbid and/or absorbing media. Fibre-optic sensors (optrodes) measure changes in absorbance of a pH-sensitive dye or changes in dye fluorescence (luminescence). An optrode can be constructed by embedding fluorescent molecules directly onto the end of an optical fibre (12). For better immobilization, a support matrix such as a polymer (13), porous glass prepared using the sol–gel process (also known as chemical solution deposition) (14, 15), or ceramic can be used to fix the fluorophore, but the ability of the support to allow quick contact between the fluorophore and the analyte ultimately determines the response characteristics of the sensor. Thin bonded layers have been used, but the thickness of a fluorophore film is crucial. Thin films respond quickly, but the output will be low because of finite molar absorptivities with limited fluorophore. Conversely, thick films give a better output signal, but the response is compromised (16). The expansion of dynamic pH ranges by using multiple pH indicators or one indicator with multiple steps of acid dissociation has been proposed (15), but range and degradation of the fluorophore remain a concern.

3.3. Ion-Selective Field-Effect Transistors

Different versions of the ISFET have been developed since the 1970s. The source and drain of an ISFET are constructed in the same manner as a standard metal oxide silicone field-effect transistor (MOSFET). In fact, the only difference between an ISFET and a MOSFET is that the gate electrode is separated from the chip and replaced by a reference electrode in contact with the gate oxide via the aqueous sample solution (17, 18), as illustrated in Fig. 5. The gain of the transistor is dependent on the concentration of ions in solution. Sensitivity is controlled by the properties of the electrolyte/insulator interface. Commonly adopted pH-sensitive layers are SiO_2, Si_3N_4, Al_2O_3, and Ta_2O_5, but an amorphous diamond-like

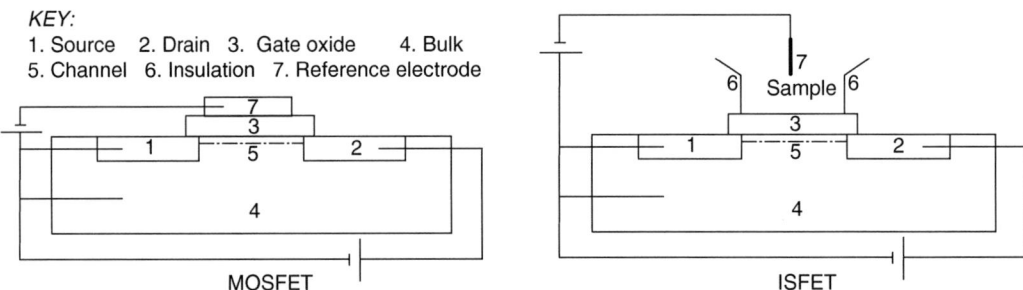

KEY:
1. Source   2. Drain   3. Gate oxide   4. Bulk
5. Channel   6. Insulation   7. Reference electrode

MOSFET   ISFET

Fig. 5. Schematic structures of MOSFET and ISFET. The difference between MOSFET and ISFET is separation of the gate electrode from the chip which is replaced by a reference electrode in contact with the gate oxide via the aqueous sample solution.

carbon (DLC) (19) has also been used as the pH-sensitive layer and passivation layer in the same application. The principles of ISFET sensors have been reviewed in detail elsewhere (17, 18).

4. Dissolved Oxygen and Dissolved Carbon Dioxide

Oxygen is typically measured as dissolved oxygen in a liquid, using a sterilizable Clark-type electrode (20). Other methods measure the amount of oxygen in exit gases and oxygen uptake rates (OURs) are typically measured in this manner. More recently, quenched fluorescence sensors have been developed and, as with pH, ISFET technology provides a promising solid-state sensor option.

Dissolved CO_2 is not as commonly measured in yeast work, but it is important in some animal cell work, where CO_2 supplementation can be necessary (21). The electrochemical approach to measuring dissolved CO_2 is basically a pH Severinghaus-type sensor with the electrode surrounded by carbonate buffer and separated from the sample by a PTFE membrane that is permeable to CO_2. A major drawback for consideration with online use is that the buffer cannot be sterilized in situ, and the response time and range are also limited. A more common measurement of CO_2 used in yeast work is off-gas analysis. Gas analysis is most commonly carried out via mass spectrometry or gas analyzers (22, 23). When the carbon dioxide evolution rate (CER) is divided by the OUR, a value for the respiratory quotient (RQ) is obtained. RQ is a useful indicator of the metabolic state of yeast growth, particularly where the Crabtree effect is of concern, such as during baker's yeast production.

The Clark electrode for dissolved oxygen measurement was first developed to measure oxygen levels in blood, but remains the most common technique for measuring dissolved oxygen in bioreactors. Inside the probe there is a platinum electrode (cathode) and a reference Ag/AgCl electrode (anode) in contact with a half saturated solution of KCl. The tip of the probe is a polypropylene or Teflon membrane through which oxygen transfers by diffusion. The platinum cathode is placed close to the membrane and reduces oxygen electrolytically, creating a current between the anode and cathode that is proportional to the oxygen partial pressure in the external broth. A very specific voltage is used between the anode and cathode to ensure only oxygen is reduced. An advantage of the Clark electrode is the linear relationship between oxygen concentration and electrical current. This means that only a two-point calibration between 0 and 100% air saturation is required to measure oxygen saturation levels in a sample. Since only the dissolved partial pressure of oxygen is measured, it is worth noting that the dissolved oxygen concentration can only be calculated if the solubility of oxygen in the broth is known at the temperature and pressure of measurement. A drawback of this sensor is that oxygen consumption at the membrane interface is equal to the rate of diffusion in the sensor. Clark electrodes should not be used in environments where oxygen levels at the permeable membrane interface are not quickly replenished. Oxygen levels must be continually replenished at the permeable membrane to ensure a reasonable response (24). Miniaturizing the sensor tip is one means of limiting oxygen consumption and improves their use in these environments (24).

Fibre-optic sensors exist for both dissolved oxygen and carbon dioxide measurement. Carbon dioxide, like pH, can be measured via changes in absorption or fluorescence of a pH-sensitive indicator. A "wet" optical sensor can be formed by placing an alkaline buffer, which contains the dye, behind a hydrophobic membrane to prevent leaching, but the preferred method is to use ion pairing which allows a "dry" sensor to be constructed (25, 26). This approach allows a non-polar indicator dye, such as hydroxypyrenetrisulfonic acid (HPTS), and a quaternary ammonium base (the ion pairing agent) to be immobilized into a gas permeable, non-polar, polymer matrix. The quaternary ammonium base also serves as the buffer (25).

Fibre-optic oxygen sensors use quenching of a fluorescent organometallic dye, usually a ruthenium (II) complex immobilized on polymeric sol–gel material or silica gel particles. Xiong and coworkers (27) report an immobilization technique using the electroless composite plating method to immobilize silica particles, absorbed with the ruthenium complex, onto a copper mesh for improved function in harsh environments. Fluorescence is at a maximum when there is no oxygen present, and thus the sensor is more sensitive at low oxygen concentration (<20%). Nevertheless, these sensors can be used in the range from 0 to 100% saturation in water and calibration can be carried out in the same manner as a Clark electrode despite their non-linear sensitivity. No oxygen is consumed, so sensitivity to stirring is minimal once the sensor is stabilized in solution.

The ISFET type sensor, described for pH measurements, can easily be adapted to sense other chemical species, such as dissolved oxygen. A two-stage process is used, such that the transducer reacts with the analyte, reducing O_2 to OH^-, and produces a pH change measurable by a pH-ISFET (28). ISFET sensors are equally adaptable for dissolved CO_2 measurement by coating a pH-ISFET with a semi-permeable membrane through which the analyte must pass to yield a pH change (29). Unfortunately, as both the O_2-ISFET and CO_2-ISFET are using concepts from their electrochemical counterparts, they suffer from consuming oxygen and from having a slow response time with drift (21).

5. Biomass Concentration

5.1. Optical Cell Density

It is generally agreed that direct measurement of turbidity remains the most common practice for estimating biomass (30). The main advantage of this technique is that it is fast and easy to perform (31, 32). The emergence of fibre optics, developed for the telecommunications industry, allows traditional optical density measurements to be taken in situ via a fibre optic probe and be delivered to a spectrometer. Charge-coupled devices (CCDs) or photodiode arrays have also replaced expensive photomultipliers in conventional spectrometers, making the technology more affordable and offering improved range into the near-infrared region.

The progress of online measurement has been regularly reviewed (30–36). Within these reviews a number of manufacturers for both probes and spectrometers are highlighted, and this is perhaps a further advantage of optical cell density measurements, as equipment can for the most part be mixed and matched, providing a customizable system. In particular, autoclavable fibre optic probes that withstand cleaning-in-place (CIP) and sterilization-in-place (SIP), fitting standard ports, are commercially available from a number of manufacturers, including ASR, Finnesse, Hellma, Optek-Danulat, and Wedgewood. Most commercial manufacturers also offer probes with a range of path lengths, over which the attenuation of light is measured.

Decisions on path length are predominantly based on the desired range and precision of measurement required. Larger path lengths are preferred for low density measurements as the larger sampling volume gives better sensitivity. Smaller path lengths are preferred for high cell density measurements, as they extend the working range. However, in certain systems the choice of path length cannot be entirely based on range. Although fibre optic probes satisfy many criteria for in situ use, drawbacks are interference from bubbles in well-aerated agitated systems and probe fouling. Increasing path length is sometimes considered useful in reducing fouling, as smaller path lengths are more prone to clogging and fouling. Nevertheless, selecting a probe constructed of smooth materials (with a surface roughness, Ra, of less than 0.4 μm) and placing probes in a high shear region (37) are recommended. Bubble interference can also be reduced to some degree with smaller path lengths, but such interference cannot be overcome to any satisfactory degree in well-aerated systems with well-dispersed bubbles. Bubble size distributions in agitated aerated vessels obtained using computational fluid dynamics (38), capillary suction (39–41), phase Doppler anemometry (42), and digital imaging (39) show that over half of bubbles in an aerated vessels, where agitation is used to disperse gases, have a diameter less than 0.5 mm and are thus smaller than the path length of commercial probes. In an air-water system, assessed using digital imaging, it was shown that 40% of all bubbles were between 0.04 and 0.3 mm in diameter (43). In a comparative assessment of bubble size studies, it was further shown that bubble size differences were not related to the measuring technique (44).

One commercial probe that attempts to overcome bubble interference is the Cerex MAX (30, 45). During operation, a Teflon plunger with an embedded magnet can be moved to open and close the side port of a sample chamber within the probe. Typically, the port is left open for 1 min to allow sample in and then closed for 1 min while gas is released from the open top port. The problem is that readings are only available once every 2 min (45) and users report problems with probe performance due to inconsistencies during manufacturing, mechanical reliability of the valve plunger and signal drift (30).

We recently developed a flow-through bubble excluder (WO/2007/096626 A1) that constantly replenishes the measurement area with sample via impeller-driven fluid flow. The excluder provides a controlled hydrodynamic environment around the measuring region of the probe by providing a calculated amount of resistance to flow across the device between inlet and outlet. The device is adjustable to any system dynamics and is simply scaled to impeller tip speed. The inlet and outlet arrangements both provide measures to prevent the ingress of bubbles.

In control experiments, varying air flowrates and agitation speeds in a solution of 0.15 M NaCl showed no interference from bubbles into the measuring window for the full range of aeration and agitation capable in the bioreactor, 0–80 L/min and 0–800 rpm, respectively (Fig. 6).

Comparison of data for standard and bubble-excluded growth of *Saccharomyces cerevisiae* W303-1A demonstrates the dual nature of bubble interference (Fig. 7). Where the concentration of bubbles is high relative to the number of cells, bubbles cause scatter that falsely increases the apparent OD by up to 1 OD unit. As the concentration of cells increases through growth, the bubbles not only scatter light, but also displace large numbers of cells that would have scattered light more effectively than the bubbles. At the mid-point

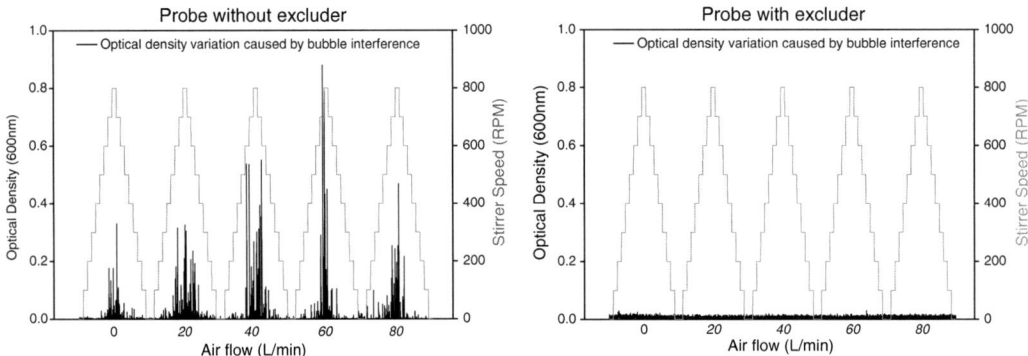

Fig. 6. Effective exclusion of air bubbles from the optical density measurements. For different air flow-rates the stirrer speed was varied stepwise in an Applikon ADI 1075, 70 L bioreactor.

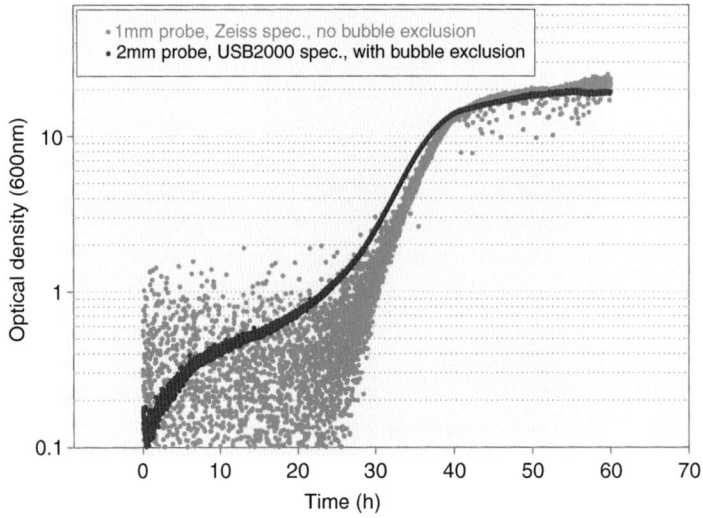

Fig. 7. Yeast growth monitored simultaneously with and without bubble exclusion.

of growth, a region is reached where the positive interference from bubbles is counteracted by the negative interference from cells being displaced, meaning that the apparent effect of this dual interference is almost negligible. As the cell concentration increases further, however, bubbles entering the measurement area displace increasingly larger numbers of cells, leading to an underestimation of the optical density in excess of 10 OD units.

Measurements made with bubble exclusion allow this technique to measure accurately across the range of a system and shows good agreement with off-line measurements. Online, in situ measurements have been used to indicate induction conditions accurately and were sensitive enough to show a difference in doubling time, as a

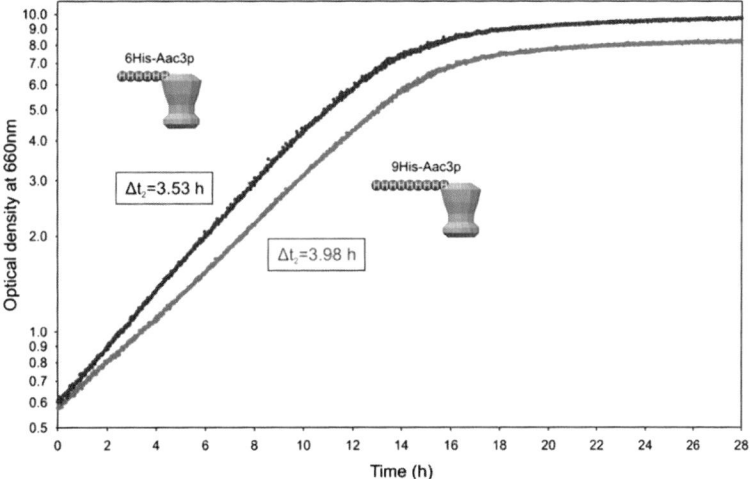

Fig. 8. Growth measurements in a fully aerated bioreactor for two strains of yeast producing the mitochondrial ADP/ATP carrier with either an amino-terminal hexa- or nona-histidine tag. Continuous measurements were carried out with a fibre-optic probe fitted with an air bubble excluder. The results show the subtle effect on the doubling time, as a consequence of the introduction of three extra histidine residues on the recombinant polyhistidine-tagged protein.

consequence of the introduction of three extra histidine residues on the recombinant histidine-tagged protein, the mitochondrial ADP/ATP carrier (Fig. 8).

5.2. Dielectric Spectroscopy (Capacitance Method)

Dielectric spectroscopy is an attractive measurement of biomass, because only living cells are measured, whereas optical density measurements cannot distinguish between live and dead cells. The theory behind this technique was first published over 50 years ago (46), and has been suitably reviewed (47, 48). Harris et al. (49) were the first to propose dielectric capacitance as a technique for online biomass monitoring and Aber Instruments Biomass Monitors have since become commonplace in research laboratories and breweries. Although most of the literature available on this technique uses Aber instruments, Hewlett-Packard (50) and Fogale Nanotech (32) have also produced analyzers.

When an electric field is applied to a suspension of cells in an aqueous ionic solution, positive and negative ions are forced to move in opposite directions in the field. As the ions reach the suspended cells, they can move no further, and thus polarization at the poles of each intact cell membrane occurs. The intact cells are now acting as electrical insulators or dielectrics. The frequency of the applied electric field has a distinct effect on the capacitance of a cell suspension, as illustrated in Fig. 9, because a finite amount of time is required for the ions to reach the cell membranes to polarize them. At low frequency, below 0.1 MHz, a single direction of ion

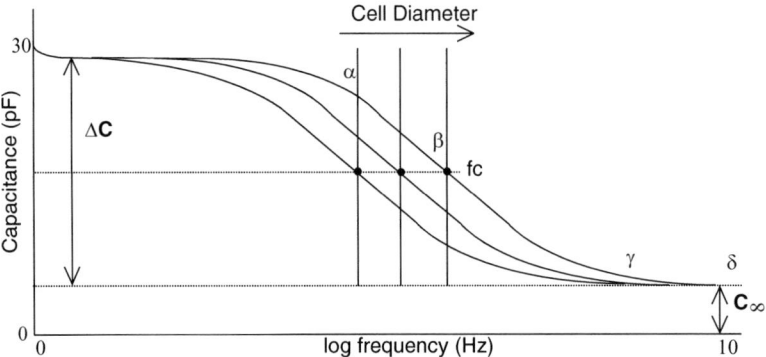

Fig. 9. Capacitance of a cell suspension as a function of frequency. At low frequency a high level of polarization is achieved. With increased frequency, fewer ions reach the cell plasma membrane so polarization and therefore capacitance are reduced. The loss of polarization can be attributed to 4 regions; α, β, γ, and δ dispersions. Cell membrane polarization is attributed to the β-dispersion. Cell capacitance is often measured at the critical frequency (f_c), where 50% polarization occurs.

motion is maintained for long enough to ensure a large degree of polarization before the field is reversed, resulting in a high capacitance reading. At high frequency, above 10 MHz, very few ions reach the plasma membrane and polarization (capacitance) is low.

The fall in capacitance as frequency increases generally occurs as a series of four step changes called dispersions (α, β, γ and δ), although these steps are not usually obvious when plotted from real data and appear smooth as illustrated in Fig. 9. α-dispersions are due to the tangential flow of ions across cell surfaces. The build-up of charge at cell membranes is attributed to the β-dispersion. γ-dispersions are due to the dipolar rotation of small molecules, particularly water. δ-dispersions are produced by the rotation of macromolecular side-chains and bound water (47, 51).

Although biomass concentration can be measured using a single frequency, normal practice is to measure capacitance at two frequencies for yeast biomass monitoring, typically 580 kHz and 15.65 MHz with the Aber Instruments Biomass Monitor. The ideal measuring frequency for any cell is dependent on morphology, size and type, but the aim is to take capacitance measurements for biomass at the critical frequency, (f_c), where 50% polarization occurs $(C_\infty+(\Delta C/2))$. The critical frequency range is typically between 0.5 and 3.0 MHz for most cell types. As already explained, frequencies above 10 MHz achieve negligible cell polarization, and this second measurement is therefore useful as a correction to compensate for the suspension medium and environmental artefacts.

Of course, products of metabolism and additions, such as pH control, all affect the background capacitance, so high frequency measurements are useful. Unfortunately, background capacitance is not the only interference, because bubbles and probe fouling are also an issue. Although bubbles have no membrane to polarize, they potentially affect capacitance measurements in the same way as optical measurements at high cell density by displacing cells that would otherwise be polarized. Fouling of the probes is also an issue and the electrode metal, which is in contact with the medium, can also become polarized (49). Fouling is dealt with through electronic (10 V) cleaning pulses applied to the electrode that generate gas bubbles by electrolysis, which lift off any adhered materials (48). Early electrode designs were more prone to polarization issues than modern electrodes because the original 4-pin probe has been replaced by electrodes assembled as rings or by a 4-pin configuration with extended flush pins. Sensitivity of capacitance measurements remains an issue at low cell concentrations, particularly if medium conductivity is high.

In advanced applications, Maskow et al. (52) have shown that lipid content in yeast could be correlated to a shift in the ratio of β-dispersion to dry cell mass. Since the lipid droplets may block intracellular ion paths, overproduced enzymes may be detectable in the same manner. Other investigators (53, 54) used dielectric spectroscopy to monitor cell cycle progression and they have shown that growth phases of a *S. cerevisiae* mutant could be detected. Results were confirmed with simultaneous measurement of DNA levels by cytometric testing (53), considering cell shape and the effect of the cell wall (54). This type of information has the potential to allow controlled substrate additions or cell harvest at specific growth stages.

5.3. In Situ Microscopy and Flow Cytometry

In situ microscopy has also been suggested for biomass monitoring (55–57), but there are concerns over online monitoring and control that include acquisition times, sterility, and applicability to small volumes (21). Microscopy does have the advantage of providing an in situ study of morphology.

In situ flow cytometry has also emerged in recent years (58, 59), but questions remain over its applicability. Reported problems include clogging and potential complications with the optimization of staining procedures (52). These measurement devices might be difficult to use because of their sophistication.

6. Other Measurements

6.1. Electrochemical Biosensors

Electrochemical biosensors have been around for decades, but their general success in providing quick and specific analyte monitoring has been questioned in a review focused on glucose sensors (60). Favourable comments are made on one blood glucose monitoring technique developed by yellow spring instruments (YSI). YSI developed a glucose sensor based on the work of Clark (20) who measured a decrease in oxygen tension resulting from the enzymatic oxidation of glucose. The unique aspect of the YSI instrument was the immobilized enzyme membrane that could be held in close proximity to the silver-platinum sensor probe within the probe holder of the device. The original YSI 23A first became popular in hospital laboratories in 1975 as a reliable blood glucose monitor, but later models such as the 2700 and 2730 are now being marketed as broad biochemistry analyzers for applications in biotechnology. Slight adjustment of the standard 2700 model, namely an external sampling pump to draw sample to the sampling station, allows the monitor to be used as an ex situ, but online, biochemistry analyzer. Each YSI 2700/2730 model has two measuring channels with an independent sensor probe. Since several different immobilized enzyme membranes are now available, the possibility exists to measure simultaneously two chemistries online, provided that the buffers used are compatible. The enzymatic oxidation reactions that are currently available through the YSI catalogue of enzyme-immobilized membranes are listed in Table 1.

The 2730 can be configured for online ex situ measurement by connecting the side-mounted pump inlet to a sampling probe, such as the Flownamics FISP probes, which have the added advantage of an in-built sample filter for retention of biomass. The 2730 can be set to draw samples at a given interval into an external station, where the sipper will pick up the sample and fill the sampling chamber of the sensor probes. The measurement time for different analytes is approximately 30 s, but a reasonable set-point can be maintained, either on a replenishment or dilutant basis through the three discrete transistor logic signals that can be used to operate a pumped addition line. The set-up used for control with a YSI 2730 is shown in Fig. 10. The sampling system is rinsed with sterilizing fluid between each reading to maintain an aseptic barrier. For accurate control, it helps to know what the maximum rate of change of the analyte to be controlled is likely to be. From this value the suitable minimum stock concentration for the feed line can be calculated. Once a stock concentration is decided, the user inputs a time-per-unit error into the YSI 2730, which is the calculated time the pump must be activated at its calibrated delivery rate to change a unit volume in the reactor by one chosen unit of mass for a chosen stock solution. In unpublished work, we have successfully maintained a glucose set-point of 0.4% w/v with good stability over a period of 72 h without contamination. The YSI 2730 can also be controlled via supervisory control and data acquisition (SCADA) software.

6.2. Fibre Optic Biosensors

Measurement of the interaction of light with a process environment is a non-invasive, non-destructive technique that offers the possibility to obtain continuous data for multiple analytes simultaneously (61). Since there are many optical measuring techniques available,

Table 1
Analytes and corresponding enzyme reactions that can be monitored by the YSI 2700/2730 electrochemical biosensors

Analyte	Immobilized enzyme reaction
D-glucose	$D\text{-glucose} + O_2 \xrightarrow{\text{gluoxidase}} H_2O_2 + \text{glucono} - \delta - \text{lactone}$
L-lactate	$L\text{-lactate} + O_2 \xrightarrow{\text{L-lacoxidase}} H_2O_2 + \text{ pyruvate}$
sucrose	$\text{sucrose} + H_2O \xrightarrow{\text{invertase}} \alpha - D\text{-glucose} + (\text{fructose})$ $\alpha - D\text{-glucose} \xleftarrow{\text{mutarotase}} \beta - D\text{-glucose}$ $\beta - D\text{-glucose} + O_2 \xrightarrow{\text{gluoxidase}} H_2O_2 + \text{glucono} - \delta - \text{lactone}$
lactose	$\text{lactose} + O_2 \xrightarrow{\text{galoxidase}} H_2O_2 + \text{galactose-dialdehyde derivative}$
ethanol	$\text{ethanol} + O_2 \xrightarrow{\text{alcohol oxidase}} H_2O_2 + \text{acetaldehyde}$
starch	$\text{starch} + H_2O \xrightarrow{\text{amyloglucosidase}} \beta - D\text{-glucose}$ $D\text{-glucose} + O_2 \xrightarrow{\text{gluoxidase}} H_2O_2 + \text{glucono} - \delta - \text{lactone}$
choline	$\text{choline} + 2O_2 \xrightarrow{\text{choloxidase}} H_2O_2 + \text{betaine}$
galactose	$\text{galactose} + O_2 \xrightarrow{\text{galoxidase}} H_2O_2 + \text{galactose-dialdehyde derivative}$
hydrogen peroxide	$H_2O_2 \xrightarrow{\text{platinum anode}} 2H^+ + O_2 + 2e^-$
L-glutamate	$L\text{-glutamate} + O_2 \xrightarrow{\text{glut oxidase}} H_2O_2 + \alpha - \text{ketogluterate} + NH_3$
L-glutamine	$L\text{-glutamine} \xrightarrow{\text{glutaminase}} L\text{-glutamate} + NH_3$ $L\text{-glutamate} + O_2 \xrightarrow{\text{glut oxidase}} H_2O_2 + \alpha - \text{ketogluterate} + NH_3$
methanol	$\text{methanol} + O_2 \xrightarrow{\text{alcohol oxidase}} H_2O_2 + \text{formaldehyde}$

including UV, IR, Raman, and fluorescence spectroscopy, there is no "single-box" technique for optical biosensors. In addition, many optical techniques are hindered by interference from bubbles and the possibility of bio-fouling with certain cell-lines.

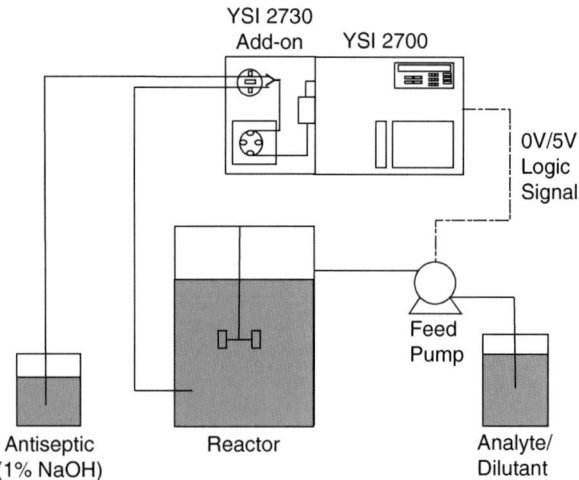

Fig. 10. Biochemistry analyzer controlled by a YSI 2730 electrochemical biosensor. A sample is pumped to a sampling station of the YSI2730 (ex situ). The analyte concentration is measured, and an output signal (0 V/+5 V Transistor Logic) is sent to an external pump to activate appropriate feeding based on the concentration and pump feed-rate of the analyte.

There are two classifications for online fibre optic sensors, direct and indirect. Direct sensors measure the intrinsic properties of the sample, for example the measurement of absorbance for biomass monitoring, including measurements of emission or refractive index. Indirect measurements include colour change caused by an immobilized fluorescent indicator, label or enzyme, such as the optical techniques already described for pH, dissolved oxygen and carbon dioxide. Modern instruments have wide enough ranges to overlap with shorter wavelengths in the near-IR range and commonly measure wavelengths up to 1,000 nm.

6.2.1. Infrared Spectroscopy

Near infra-red spectroscopy (NIRS) and mid infra-red spectroscopy (MIRS) spectroscopy are useful tools for monitoring organic systems because bonds, such as C–H, N–H, and O–H all absorb electromagnetic energy in this range (700–2,500 nm). NIRS is generally preferred to MIRS because NIRS allows direct analysis without any sample preparation (63), even though NIRS bands are 10–100 times weaker than those for MIRS (62).

Position, shape, and size of an absorption band can be related to specific analytes. However, absorption bands are broad and overlapping, and since most analytes absorb at more than one wavelength each band includes contributions from more than one analyte. Suggested ranges for the measurement of specific analytes vary greatly in the literature and the optimal wavelength for the measurement of an analyte online and at-line also differ (32).

NIRS has been used to measure glucose, fructose, glutamate, glutamine, proline, ammonia, carbon dioxide, and phosphate concentrations (64). Nevertheless, interpretation of spectral data is challenging and time consuming (21, 65) and most users report difficulties in calibration and interference as the main challenges. For interpretation, multivariate regression such as partial least squares (PLS) analysis, principle component analysis (PCA) or neural networks is required to calibrate IR-spectra with reference analytical methods. A preprocessing digital-Fourier-filtering step can also improve results (66).

Cavinato and co-workers (67) used NIRS between 700 and 1,100 nm to monitor the ethanol concentration through the glass wall of a bioreactor. They were able to apply a successful calibration model, but highlight the inability to measure trace levels of ethanol compared to gas chromatography, flow injection analysis, and enzyme-based techniques. Finn, Harvey, and McNeil (68) used NIRS to monitor biomass, ethanol, glucose, and protein content simultaneously in high density fed-batch baker's yeast production. These investigators describe the model-building process for each analyte in a challenging system where there is a high level of light scattering from the cell mass but relatively low concentrations of analyte. These workers also used a chemically defined medium that avoids the additional complication of factoring a complex, ill-defined medium into a robust calibration model (65). Arnold and co-workers (63) report that the path length of fibre optic probes used for NIRS was a critical factor in accuracy. Tamburini and co-workers (69) report that both agitation and aeration rates markedly influenced the spectral signals.

6.2.2. Fluorescence Spectroscopy

The most common application of fluorescence sensors is the measurement of NADH and NAD(P)H as a biomass concentration indicator. NAD(P)H-dependent fluorescence can be measured at 450 nm after excitation at 360 nm. Unfortunately, the correlation with biomass concentration only holds if the concentration of NAD(P)H is constant within the cell. Applications are limited to defined media that do not absorb at the emission or excitation wavelength, and to cases where the metabolism does not alter the NAD(P)H concentration. The BioView sensor (Delta Light and Optics, Denmark) is one commercially available device used by several investigators (66, 70) to measure biomass concentration. Other fluorophores that are commonly measured using fluorescence are vitamins, such as pyridoxine and riboflavin, and amino acids, such as tryptophan. Tryptophan has also been suggested as a better indicator of cell concentration than NAD(P)H (71): NAD(P)H levels inside the cell are reported to fluctuate with oxygen variation (72), as NADH is re-oxidized to NAD^+ by oxidative phosphorylation (73).

6.2.3. Raman
Spectroscopy

Raman spectroscopy measures the vibrational energies of molecules, but these vibrational energies can only be detected for molecules that exhibit a change in polarization after the collision of photons with the molecule. Raman and infrared spectroscopy are potentially complementary, but Raman spectroscopy has been used very little for bioprocess monitoring because some biological molecules fluoresce within Raman scattering band regions. More recent application of Raman spectroscopy incorporates a new technique called shifted subtracted Raman spectroscopy (SSRS) (74), also termed "scissors". This method serves to reduce noise and fluorescence by collecting sample spectra at different angles and cancelling detected irregularities. Such advances should allow more applicability to bioprocesses in the future. Examples of work that could be applied to yeast in the future include that of Lee and co-workers (75), who simultaneously monitored glucose, acetate, formate, lactate, and phenyl-alanine in a phenyl-analine-producing *E. coli* culture. FT-Raman spectroscopy (76) and dispersive Raman spectroscopy (77) have both been reported to offer high precision monitoring for glucose and ethanol data during the biotransformation of glucose to ethanol by yeast.

6.3. Reporter Genes

Because of their obvious applicability to monitoring gene expression in yeasts, reporter genes, although synonymous with both fluorescence and UV–vis spectroscopy that have been mentioned already, deserve individual mention. Reporter genes are genetically linked to the structural gene, allowing the expression of the structural gene to be monitored. The most commonly used reporter/marker for yeast is the green fluorescent protein (GFP), although the use of others, including luciferase (78) and blue fluorescent protein (BFP) (79) are also reported.

Wild-type GFP is sourced from the jellyfish *Aequorea victoria* and exhibits green fluorescent light when exposed to blue light. This wild-type GFP has dual excitation peaks at wavelengths of 395 and 475 nm and an emission peak at 509 nm. Potential downsides to tagging GFP and other markers onto a gene sequence are, firstly, that synthesis of the marker can add an extra burden to the overall metabolism of the cell and, secondly, that the fluorescent signal can lag behind expression by as much as 95 min (80, 81). Fortunately, both factors have been minimized in new variants. Cormack and co-workers (82) created a library of mutant GFP genes in *E. coli* with random amino acid substitutions and used fluorescence-activated cell sorting (FACS) to select variants of GFP with increased fluorescence intensity, 20–35 times that of the wild-type. They report that the gain through selection of intensity, coupled with better protein folding of the mutant GFP over wild-type in *E. coli*, resulted in a 100-fold increase in fluorescence intensity. Lag times of less than 25 min are reported in this work. Some workers, using Gram-negative bacteria report lag times as little as

8 min (83). In most cases, the culture time of yeast is significantly longer than that for bacteria, and thus the lag time for chromophore formation has less impact on the real-time protein production estimation (81).

6.4. Electronic Tongues

At present, there is little evidence of electronic tongues providing an online, in situ option for bioprocess monitoring. Nevertheless, the argument has been made that they could one day be applied more readily, as the cross sensitive sensors used in electronic tongues are often composed of well-known autoclavable sensors (21).

Electronic tongues are generally made up of an array of non-specific sensors, such as amperometric, potentiometric, or optical sensors, to produce a fingerprint of the measured sample through analysis of multi-component matrices. The common analogy of the electronic tongue is that of biological organization of the olfactory and taste system in mammals. There, dozens of receptors send signals to the brain to be processed by nets of neurons and provide a sensed image of the object. Signal processing of an electronic tongue relies on complicated mathematical procedures based on pattern recognition (PARC) and other multivariate analysis techniques, such as artificial neural networks (ANN), or PCA. Several workers have carried out analysis of culture media (84–86) to prove the potential of electronic tongues as a measurement and control tool, but the advancement towards in situ applications will decide its future potential.

7. Process Control

In most commercial cultures, parameters such as temperature, pH, aeration, agitation, and dissolved oxygen are measured and controlled via the feedback approach outlined in Fig. 11. Basic culture variables can be maintained at predetermined levels through actuators, such as potentiometers, to control stirrer motors, pneumatic valves, pumps, and mass flow controllers (MFC) to vary the supply of liquids and gases. The measured values of the controlled parameter are signaled to the controller, where the value is compared to set-point values. The controller calculates the error value, which is used by the controller to action the appropriate corrective measure through the actuator. For different variables, the type of actuator and the necessary level of control are dependent on the nature of the specific variable.

For pH control, digital on–off control using a single speed pump as the actuator is usually sufficient, although a small level of deviation is usually tolerated (typically ± 0.1) to avoid problems of rapid switching and measurement delay. Other parameters, such as dissolved oxygen, require a more sophisticated approach to control

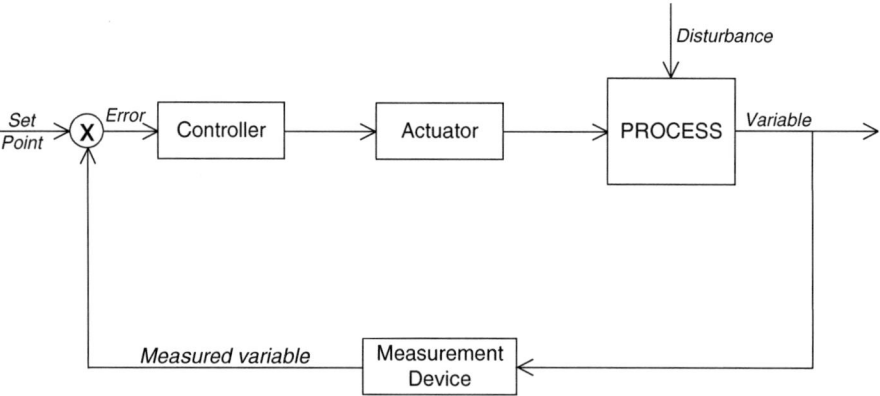

Fig. 11. Stages of a feedback control loop. A measuring device relays the measured process variable back to a controller. The controller calculates the "strength" of the control action required and actuates a response through specific actuators, such as valves, variable MFC, potentiometers and pumps. This manipulation, plus any parallel process disturbance, is sensed in the next measurement of the continuous loop.

and analogue signals are fed into variable MFC on gas inlet lines. Potentiometers are used to vary impeller speed, such that the "strength" of the control action can be varied proportional-integral (PI) control, probably the most common type of control used for basic bioprocesses, and proportional-integral-derivative (PID) control provide a more intuitive means of determining the control action from a combination of two or three linear functions of the error signal, respectively. The proportional component (P) sets the scale of reaction based on current error from set-point, whereas the integral function (I) determines a further output function based on the sum of recent errors. The derivative function (D) is used to determine a reaction based on the rate at which an error has been changing. Although PI and PID control is sufficient to control most variables in a bioprocess, it is imperative that the constants, P, I, and D of the controller algorithm are tuned correctly. A poorly tuned controller will invariably result in unstable process control.

Although PID controllers are considered to be applicable for most process control, they are not adaptive controllers and do not learn from previous errors. As such, they remain most suited to relatively simple control tasks (87), and are preferred for control of linear process behaviour or, at least, for process behaviour that is pseudo-linear by appearing linear over finite regions, even if the process is non-linear over a larger region (88). Advanced methods of control include model predictive control, neural networks, fuzzy logic, or some combination of these. Hybrid models, combining mathematical models, historical data, live data, and human expertise are considered especially useful for complicated applications (89).

The mathematical expenditure required for advanced control is always increased (90), but neural networks and fuzzy control both have an inherent ability to handle non-linear relationships so, in cases where PID control is insufficient, these options are considered useful. Fed-batch systems, frequently used to avoid substrate inhibition, are a popular case study for advanced control methods (91) because they carry the additional problem of a dynamic feed-rate that has to be configured into a process model already containing many other variable, non-linear, parameters.

Jin and co-workers (92) used two ANN to estimate glucose and galactose concentrations from online measurements of ethanol, biomass, culture volume, and the amount of carbon source fed to the bioreactor, to optimize recombinant protein production. Cos and co-workers (93) coupled a predictive control algorithm to a feedback PI controller to optimize the feed rate of methanol for heterologous protein production in *Pichia pastoris*. Hisbullah and co-workers (94) found some oscillation in substrate concentration using a fuzzy logic controller for regulating substrate feed during Baker's yeast production. These oscillations were resolved using a hybrid system involving adaptive fuzzy control coupled to a scheduled-gain PI controller.

The progress of process modelling and control has been reviewed in more detail by several others (87, 88, 95, 96).

8. Conclusions

The amount of research focused on bioprocess analysis and control is not trivial but, relative to the enormous scope and complexity of biological systems, there are very few established online, in situ, measurements available as direct control parameters. A major factor in the success of any proposed online in situ sensor is its immunity to harsh conditions over extended time periods. Few of the emerging techniques highlighted in this review can claim to stand up to all prerequisites of bioprocess control concerning lifespan, range, signal drift, interference, and fouling. Even accepted measuring techniques have disadvantages attached to them that are more often managed than resolved.

New measurements are always incorporated slowly, but the pre-existence of an acceptable technique, as with pH and dissolved oxygen, can also hinder the progress of new technology into the mainstream. Despite this, a number of promising advances have appeared in the literature and commercial examples have been quoted herein.

Feasible options, which improve on existing sensors, appear to come mainly from the solid-state category with sensors for both pH and dissolved gases available commercially. The emergence of wide ranging optical methods that can provide a wealth of process information also looks promising and data-evaluation packages are constantly being developed to aid interpretation.

Optical methods have the advantage of being based on techniques with a proven robustness, rapid response, good sensitivity, and ease of maintenance, so their wider acceptance seems likely in the future.

The use of multi-sensor devices such as the electronic tongue is also interesting, but future application will rely on progress in analysis tools to aid interpretation, and work towards proving their in situ capability.

As new measurements become available and expand the possibilities for control, the need for advanced control models will grow in tandem. Simple PID control is suitable for many mainstream physical parameters, but broader application of more intuitive control is inevitable in order to deal with more complicated systems and multivariate models.

References

1. Wang TP, Wells A, Bediones D (1989) Accuracy and repeatability of temperature measurement by RTDS. Adv Instrument Control 44:1571–88

2. Sulciner JD (1996) Understanding and using PRTD technology part I: history, principles, and designs. Sensors 13(8)

3. O'Hare D, Parker KH, Winlove CP (2006) Metal-metal oxide pH sensors for physiological application. Med Eng Phys 28:982–8

4. Yuqing M, Jianrong C, Keming F (2005) New technology for the detection of pH. J Biochem Biophys Methods 63:1–9

5. Burke LD, Mulcahy JK, Whelan DP (1984) Preparation of an oxidized iridium electrode and the variation of its potential with pH. J Electroanal Chem 163:117–128

6. Martinez CCM, Madrid RE, Felice CJ (2009) A pH sensor based on a stainless steel electrode electrodeposited with iridium oxide. IEEE Trans Edu 52:133–136

7. Heather AE, Christopher FM, Marcin M (2009) Effects of electrodeposition conditions and protocol on the properties of iridium oxide pH sensor electrodes. J Electrochem Soc 156:F1–F6

8. Katsube T, Lauks I, Zemel JN (1982) pH sensitive sputtered iridium oxide-films. Sensors Actuators 2:399–410

9. Ardizzone S, Carugati A, Trasatti S (1981) Properties of thermally prepared iridium dioxide electrodes. J Electroanal Chem 126:287–292

10. Wang M, Yao S, Madou M (2002) A long-term stable iridium oxide pH electrode. Sensors Actuators B 81:313–315

11. Yao S, Wang M, Madou M (2001) A pH electrode based on melt-oxidized iridium oxide. J Electrochem Soc 148:H29–H36

12. Jin Z, Su Y, Duan Y (2000) Improved optical pH sensor based on polyaniline. Sensors Actuators B 71:118–122

13. Jones TP, Porter MD (1988) Optical pH sensor based on the chemical modification of a porous polymer film. Anal Chem 60:404–406

14. Hench LL, West JK (1990) The sol–gel process. Chem Rev 90:33–72

15. Lin J, Liu D (2000) An optical pH sensor with a linear response over a broad range. Anal Chim Acta 408:49–55

16. Grummt UW, Pron A, Zagorska M, Lefrant S (1997) Polyaniline based optical pH sensor. Anal Chim Acta 357:253–259

17. Bergveld P (2003) Thirty years of ISFETOLOGY: what happened in the past 30 years and what may happen in the next 30 years. Sensors Actuators B 88:1–20

18. Yuqing M, Jianguo G, Jianrong C (2003) Ion sensitive field effect transducer-based biosensors. Biotechnol Adv 21:527–34

19. Voigt H, Schitthelm F, Lange T, Kullick T, Ferretti R (1997) Diamond-like carbon-gate pH-ISFET. Sensors Actuators B 44:441–445

20. Johnson MJ, Borkowski J, Engblom C (2000) Steam sterilizable probes for dissolved oxygen measurement (Reprinted from Biotechnol Bioeng, 6, 457, 1964). Biotechnol Bioeng 67:645–656

21. Vojinovic V, Cabral JMS, Fonseca LP (2006) Real-time bioprocess monitoring part I: in situ sensors. Sensors Actuators B 114:1083–1091

22. Xiong ZQ, Guo MJ, Guo YX, Chu J, Zhuang YP, Wang NS, Zhang SL. RQ feedback control for simultaneous improvement of GSH yield and GSH content in *Saccharomyces cerevisiae* T65. Enzyme and Microbial Technology 46:598–602

23. Wu L, Lange HC, Van Gulik WM, Heijnen JJ (2003) Determination of in vivo oxygen uptake and carbon dioxide evolution rates from off-gas measurements under highly dynamic conditions. Biotechnol Bioeng 81:448–458

24. Doran PM (1995) Bioprocess engineering principles. Academic Press 205–206

25. Burke CS, Markey A, Nooney RI, Byme P, McDonagh C (2006) Development of an optical sensor probe for the detection of dissolved carbon dioxide. Sensors Actuators B 119:288–294

26. Weigl BH, Wolfbeis OS (1995) New hydrophobic materials for optical carbon-dioxide sensors based on ion-pairing. Anal Chim Acta 302:249–254

27. Xiong XL, Xiao D, Choi MMF (2006) Dissolved oxygen sensor based on fluorescence quenching of oxygen-sensitive ruthenium complex immobilized on silica–Ni–P composite coating. Sensors Actuators B 117:172–176

28. Sohn BK, Kim CS (1996) A new pH-ISFET based dissolved oxygen sensor by employing electrolysis of oxygen. Sensors Actuators 34:435–440

29. Uttamlal M, Walt DR (1995) A fiberoptic carbon-dioxide sensor for fermentation monitoring. Bio-Technol 13:597–601

30. Junker BH, Reddy J, Gbewonyo K, Greasham R (1994) Online and in situ monitoring technology for cell density measurement in microbial and animal-cell cultures. Bioprocess Eng 10:195–207

31. Olsson L, Nielsen J (1997) On-line and in situ monitoring of biomass in submerged cultivations. Trends Biotechnol 15:517–522

32. Kiviharju K, Salonen K, Moilanen U, Eerikainen T (2008) Biomass measurement online: the performance of in situ measurements and software sensors. J Ind Microbiol Biotechnol 35:657–65

33. Mallette MF (1969) Evaluation of growth by physical and chemical means. In: Norris JR, Ribbons DW (eds) Methods in microbiology. Academic, London, England and New York, pp 521–566

34. Clarke DJ, Blakecoleman BC, Carr RJG, Calder MR, Atkinson T (1986) Monitoring reactor biomass. Trends Biotechnol 4:173–178

35. Sonnleitner B, Locher G, Fiechter A (1992) Biomass determination. J Biotechnol 25:5–22

36. Singh A, Kuhad RC, Sahai V, Ghosh P (1994) Evaluation of Biomass. Advances in *Biochem* Bioeng/Biotechnol 51:47–70

37. Wu P, Ozturk SS, Blackie JD, Thrift JC, Figueroa C, Naveh D (1995) Evaluation and applications of optical-cell density probes in mammalian-cell bioreactors. Biotechnol Bioeng 45:495–502

38. Kerdouss F, Bannari A, Proulx P (2006) CFD modeling of gas dispersion and bubble size in a double turbine stirred tank. Chem Eng Sci 61:3313–3322

39. Laakkonen M, Moilanen P, Alopaeus V, Aittamaa J (2007) Modelling local bubble size distributions in agitated vessels. Chem Eng Sci 62:721–740

40. Alves SS, Maia CI, Vasconcelos JMT (2002) Experimental and modelling study of gas dispersion in a double turbine stirred tank. Chem Eng Sci 57:487–496

41. Barigou M, Greaves M (1992) Bubble-size distributions in a mechanically agitated gas–liquid contactor. Chem Eng Sci 47:2009–2025

42. Schafer M, Wachter P, Durst F (2000) Experimental investigation of local bubble size distributions in stirred vessels using phase Doppler anemometry. Proceedings of the 10th European Conference on Mixing 205–212

43. Machon V, Pacek AW, Nienow AW (1997) Bubble sizes in electrolyte and alcohol solutions in a turbulent stirred vessel. Chem Eng Res Design 75:339–348

44. Alves SS, Maia CI, Vasconcelos JMT, Serralheiro AJ (2002) Bubble size in aerated stirred tanks. Chem Eng J 89:109–117

45. Shiloach J, Bahar S, Miller B (1994) Online monitoring of bacterial mass during production of recombinant exotoxin-A: using an in situ steam sterilizable sensor. Biochem Eng 745:244–250

46. Schwan HP (1957) Electrical properties of tissue and cell suspensions. Adv Biol Med Phys 5:147–209

47. Markx GH, Davey CL (1999) The dielectric properties of biological cells at radiofrequencies: applications in biotechnology. Enz Microb Technol 25:161–171

48. Yardley YE, Kell DB, Barrett J, Davey CL (2000) On-line, real-time measurements of cellular biomass using dielectric spectroscopy. Biotechnol Genetic Eng Rev 17:3–35

49. Harris CM, Todd RW, Bungard SJ, Lovitt RW, Morris JG, Kell DB (1987) Dielectric permittivity of microbial suspensions at radio frequencies: a novel method for the real-time estimation of microbial biomass. Enz Microb Technol 9:181–186

50. Soley A, Lecina M, Gamez X, Cairo JJ, Riu P, Rosell X, Bragos R, Godia F (2005) On-line monitoring of yeast cell growth by impedance spectroscopy. J Biotechnol 118:398–405

51. Carvell JP, Dowd JE (2006) On-line measurements and control of viable cell density in cell culture manufacturing processes using radio-frequency impedance. Cytotechnol 50:35–48

52. Maskow T, Roellich A, Fetzer I, Ackermann JU, Harms H (2008) On-line monitoring of lipid storage in yeasts using impedance spectroscopy. J Biotechnol 135:64–70

53. Asami K, Takahashi K, Shirahige K (2000) Progression of cell cycle monitored by dielectric spectroscopy and flow-cytometric analysis of DNA content. Yeast 16:1359–1363

54. Gheorghiu E, Asami K (1998) Monitoring cell cycle by impedance spectroscopy: experimental and theoretical aspects. Bioelectrochem Bioenerg 45:139–143

55. Guez JS, Cassar JP, Wartelle F, Dhulster P, Suhr H (2010) The viability of animal cell cultures in bioreactors: can it be estimated online by using in situ microscopy? Process Biochem 45:288–291

56. Frerichs JG, Joeris K, Scheper T, Konstantinov K (2001) In situ microscopy for on-line and in-line monitoring of cell populations in bioreactors. Animal Cell Technol 1:452–454

57. Bittner C, Wehnert G, Scheper T (1998) In situ microscopy for on-line determination of biomass. Biotechnol Bioeng 60:24–35

58. Kacmar J, Carlson R, Balogh SJ, Srienc F (2006) Staining and quantification of poly-3-hydroxybutyrate in *Saccharomyces cerevisiae* and *Cupriavidus necator* cell populations using automated flow cytometry. Cytometry 69A:27–35

59. Kacmar J, Zamamiri A, Carlson R, Abu-Absi NR, Srienc F (2004) Single-cell variability in growing *Saccharomyces cerevisiae* cell populations measured with automated flow cytometry. J Biotechnol 109:239–254

60. Magner E (1998) Trends in electrochemical biosensors. Analyst 123:1967–1970

61. Ulber R, Frerichs JG, Beutel S (2003) Optical sensor systems for bioprocess monitoring. Anal Bioanal Chem 376:342–348

62. Brimmer PJ, Hall JW (1993) Determination of nutrient levels in a bioprocess using near-infrared spectroscopy. Can J Appl Spectrosc 38:155–162

63. Arnold SA, Gaensakoo R, Harvey LM, McNeil B (2002) Use of at-line and in situ near-infrared spectroscopy to monitor biomass in an industrial fed-batch *Escherichia coli* process. Biotechnol Bioeng 80:405–413

64. Harms P, Kostov Y, Rao G (2002) Bioprocess monitoring. Curr Opin Biotechnol 13:124–127

65. Ferreira AP, Alves TP, Menezes JC (2005) Monitoring complex media fermentations with near-infrared spectroscopy: comparison of different variable selection methods. Biotechnol Bioeng 91:474–481

66. Marose S, Lindemann C, Ulber R, Scheper T (1999) Optical sensor systems for bioprocess monitoring. Trends Biotechnol 17:30–34

67. Cavinato AG, Mayes DM, Ge ZH, Callis JB (1990) Noninvasive method for monitoring ethanol in fermentation processes using fiberoptic near-infrared spectroscopy. Anal Chem 62:1977–1982

68. Finn B, Harvey LM, McNeil B (2006) Near-infrared spectroscopic monitoring of biomass, glucose, ethanol and protein content in a high cell density baker's yeast fed-batch bioprocess. Yeast 23:507–517

69. Tamburini E, Vaccari G, Tosi S, Trilli A (2003) Near-infrared spectroscopy: a tool for monitoring submerged fermentation processes using an immersion optical-fiber probe. Appl Spectrosc 57:132–138

70. Haack MB, Eliasson A, Olsson L (2004) On-line cell mass monitoring of *Saccharomyces cerevisiae* cultivations by multi-wavelength fluorescence. J Biotechnol 114:199–208

71. Horvath JJ, Glazier SA, Spangler CJ (1993) In situ fluorescence cell mass measurements of *Saccharomyces cerevisiae* using cellular tryptophan. Biotechnol Progr 9:666–670

72. Siano SA, Mutharasan R (1989) NADH and flavin fluorescence responses of starved yeast cultures to substrate additions. Biotechnol Bioeng 34:660–670

73. Hisiger S, Jolicoeur M (2005) A multiwavelength fluorescence probe: is one probe capable for on-line monitoring of recombinant protein production and biomass activity? J Biotechnol 117:325–336

74. Bell SEJ, Bourguignon ESO, Dennis A (1998) Analysis of luminescent samples using subtracted shifted Raman spectroscopy. Analyst 123:1729–1734

75. Lee HLT, Boccazzi P, Gorret N, Ram RJ, Sinskey AJ (2004) In situ bioprocess monitoring of *Escherichia coli* bioreactions using Raman spectroscopy. Vibr Spectrosc 35:131–137

76. Sivakesava S, Irudayaraj J, Demirci A (2001) Monitoring a bioprocess for ethanol production using FT-MIR and FT-Raman spectroscopy. J Ind Microbiol Biotechnol 26:185–190

77. Shaw AD, Kaderbhai N, Jones A, Woodward AM, Goodacre R, Rowland JJ, Kell DB (1999) Noninvasive, on-line monitoring of the biotransformation by yeast of glucose to ethanol using dispersive Raman spectroscopy and chemometrics. Appl Spectrosc 53:1419–1428

78. Funabashi H, Imajo T, Kojima J, Kobatake E, Aizawa M (1999) Bioluminescent monitoring of intracellular ATP during fermentation. Luminescence 14:291–296

79. Reischer H, Schotola I, Striedner G, Potschacher F, Bayer K (2004) Evaluation of the GFP signal and its aptitude for novel on-line monitoring strategies of recombinant fermentation processes. J Biotechnol 108:115–125

80. Albano CR, RandersEichhorn L, Chang Q, Bentley WE, Rao G (1996) Quantitative measurement of green fluorescent protein expression. Biotechnol Tech 10:953–958

81. Li JC, Wang S, VanDusen WJ, Schultz LD, George HA, Herber WK, Chae HJ, Bentley WE, Rao G (2000) Green fluorescent protein in *Saccharomyces cerevisiae*: real-time studies of the *GAL1* promoter. Biotechnol Bioeng 70:187–196

82. Cormack BP, Valdivia RH, Falkow S (1996) FACS-optimized mutants of the green fluorescent protein (GFP). Gene 173:33–38

83. Handfield M, Schweizer HP, Mahan MJ, Sanschagrin F, Hoang T, Levesque RC (1998) ASD-GFP vectors for in vivo expression technology in *Pseudomonas aeruginosa* and other gram-negative bacteria. Biotechniques 24: 261–264

84. Legin A, Kirsanov D, Rudnitskaya A, Iversen JJL, Seleznev B, Esbensen KH, Mortensen J, Houmoller LP, Vlasov Y (2004) Multicomponent analysis of fermentation growth media using the electronic tongue (ET). Talanta 64:766–772

85. Turner C, Rudnitskaya A, Legin A (2003) Monitoring batch fermentations with an electronic tongue. J Biotechnol 103:87–91

86. Esbensen K, Kirsanov D, Legin A, Rudnitskaya A, Mortensen J, Pedersen J, Vognsen L, Makarychev-Mikhailov S, Vlasov Y (2004) Fermentation monitoring using multisensor systems: feasibility study of the electronic tongue. Anal Bioanal Chem 378:391–395

87. Lubbert A, Jorgensen SB (2001) Bioreactor performance: a more scientific approach for practice. J Biotechnol 85:187–212

88. Alford JS (2006) Bioprocess control: advances and challenges. Comput ChemEng 30: 1464–1475

89. Schugerl K (2001) Progress in monitoring, modeling and control of bioprocesses during the last 20 years. J Biotechnol 85:149–173

90. Ritzka A, Sosnitza P, Ulber R, Scheper T (1997) Fermentation monitoring and process control. Curr Opin Biotechnol 8:160–164

91. Nyttle VG, Chidambaram M (1993) Fuzzy-logic control of a fed-batch fermenter. Bioprocess Eng 9:115–118

92. Jin S, Ye KM, Shimizu K, Nikawa J (1996) Application of artificial neural network and fuzzy control for fed-batch cultivation of recombinant *Saccharomyces cerevisiae*. J Ferm Bioeng 81:412–421

93. Cos O, Ramon R, Montesinos JL, Vallero F (2006) A simple model-based control for *Pichia pastoris* allows a more efficient heterologous protein production bioprocess. Biotechnol Bioeng 95:145–154

94. Hisbullah, Hussain MA, Ramachandran KB (2003) Design of a fuzzy logic controller for regulating substrate feed to fed-batch fermentation. Food Bioproducts Process 81:138–146

95. Rani KY, Rao VSR (1999) Control of fermenters: a review. Bioprocess Eng 21:77–88

96. Shioya S, Shimizu K, Yoshida T (1999) Knowledge-based design and operation of bioprocess systems. J Biosci Bioeng 87:261–266

Chapter 13

Monitoring the Biomass Accumulation of Recombinant Yeast Cultures: Offline Estimations of Dry Cell Mass and Cell Counts

Shane M. Palmer and Edmund R.S. Kunji

Abstract

Biomass is one of the most important parameters for process optimization, scale-up and control in recombinant protein production experiments. However, a standard unit of biomass remains elusive. Methods of biomass monitoring have increasingly been developed towards online, in situ techniques in order to advance process analysis and control. Offline, ex situ methods, such as dry cell mass determination and direct cell counts, remain the reference for determining cell mass and number, respectively, but this type of analysis is time consuming. In this chapter, protocols are presented for determining these offline measures of the biomass yield of recombinant yeast cultures.

Key words: Biomass, Accumulation, Monitoring, Yeast, Offline

1. Introduction

Biomass monitoring remains one of the most important, but challenging, measurements in the field of biotechnology. Decades of effort have yielded a spectrum of methods, but a diverse approach means a standard unit of biomass remains elusive. Offline, at-line and online systems have been developed, and both in situ and ex situ monitoring methods are available. Offline measurement techniques are often time consuming, but remain widely practiced as calibration standards for other methods. Dry cell weight evaluation remains the definitive measure of cell mass and provides the most meaningful measure of the cellular contents (protoplasm). Cell counts are the reference method for determining cell number, while cell viability can be evaluated in tandem when combined with staining techniques that distinguish between viable and non-viable cells.

Roslyn M. Bill (ed.), *Recombinant Protein Production in Yeast: Methods and Protocols*, Methods in Molecular Biology, vol. 866, DOI 10.1007/978-1-61779-770-5_13, © Springer Science+Business Media, LLC 2012

The major advantage with any offline technique is that the sampling and/or measuring methods do not have to withstand the system environment and procedures, such as sterilization. However, samples cannot be returned to the system and preparation must be taken to meet measurement criteria, such as instrument range and linearity of calibration, which may necessitate dilution. In this chapter, protocols are given for determining cell mass, total cell counts and viability counts.

2. Materials

2.1. Dry Cell Mass

1. Phosphate buffered saline (PBS): Prepare 10× stock with 1.37 M NaCl, 27 mM KCl, 100 mM Na_2HPO_4, 18 mM KH_2PO_4 (pH to 7.4 with HCl if necessary), autoclave and store at room temperature. Prepare a working solution by mixing one part of the stock with nine parts water (see Note 1).

2. Syringe, sized to suit the sample volume, with luer lock or luer slip fitting (Becton Dickinson, from Industrial and Scientific Supplies).

3. 25-mm syringe filter holder (Sartorius).

4. Filter paper (Whatmann GF/C, 25 mm, from VWR).

5. Falcon tubes (Sarstedt, from VWR).

6. Bench-top centrifuge, such as Kendro Biofuge Prime with #7590 rotor to accommodate Falcon tubes with the appropriate holders (Kendro).

7. Drying oven (105°C) or freeze dryer.

8. Desiccator (VWR).

9. Analytical balance (Mettler Toledo AB104-S, from Jencons).

2.2. Total Cell Count and Viability Counts

1. Haemocytometer (Fisher).

2. Haemocytometer cover-slips (Fisher).

3. Ethanol.

4. Lens-cleaning tissue (Whatman No. 105, from VWR).

5. 0.4% trypan blue stain, fresh and filtered.

6. PBS.

7. Tally counter(s).

8. Gilson pipettes (Anachem).

9. Light transmission microscope with minimum 40× objective (Nikon).

3. Methods

3.1. Dry Cell Mass

More than one method can be adopted when carrying out gravimetric techniques, so an appreciation of associated errors with each technique is useful. Major decisions when measuring dry cell mass are based around how the cell fraction will be separated, whether or not cell washing will be carried out and to what extent, and how, the cells will be dried.

Cell separation techniques are often governed by the volume of sample available and/or cell concentration. The cumulative volume of multiple samples required for time-course analysis should always be considered in relation to the size of the culture volume being monitored. A small culture volume may rule out centrifugation because a relatively large volume is removed overall (see Notes 2 and 3 for further considerations regarding separation).

It is worth noting that in cases where a less accurate, but quicker, wet weight estimation of cell mass is sufficient, centrifugation yields more reproducible water content than filtration.

3.1.1. Filtration

1. Filter papers should be pre-dried at 105°C or placed in a microwave oven at 150 W for 10 min (1). Before use, filters should be cooled to room temperature in a desiccator (see Note 4). Filter papers can be kept in the desiccator prior to use, but should be pre-weighed within 15 min of removal from the desiccator.

2. Assemble the dried filter paper into the filter holder.

3. Accurately measure an appropriate volume of sample, transfer to a syringe of suitable volume (see Note 5) and force it through the filter.

4. Wash the cells by forcing one to two times the original sample volume of PBS through the filter (see Note 6).

5. Carefully remove the filter paper from the syringe and place it in a dish in a drying oven at 105°C for 12–16 h or until a stable weight is reached (see Note 7).

6. Move the dried filter to a desiccator and allow it to reach room temperature. Once at room temperature, remove the filter from the desiccator and weigh within 15 min.

7. Subtract the pre-weight of the filter, noted in step 1, from the total weight to give the dry cell weight per starting volume.

8. Scale the weight of cells to the preferred specific volume basis.

3.1.2. Centrifugation

1. Pre-dry a centrifuge or Falcon tube at 105°C. Before use, tubes should be cooled to room temperature in a desiccator. Tubes can be kept in the desiccator prior to use, but should be pre-weighed within 15 min of removal from the desiccator.

2. Accurately measure a volume of sample, place in the pre-weighed Falcon tube (see Note 8).

3. Weigh the tube and fill another tube with water to balance the sample tube in the centrifuge.

4. Spin at $12,000 \times g$ for 15 min.

5. Remove the tube from the centrifuge, taking care not to dislodge cells from the pellet, and remove the supernatant by decanting or pipetting.

6. Resuspend the cell pellet in a volume of PBS equivalent to the original sample volume.

7. Centrifuge the washed cells at $12,000 \times g$ for 15 min.

8. Remove the tube from the centrifuge, taking care not to dislodge cells, and remove the supernatant by decanting or pipetting.

9. Repeat steps 4–6 if a second wash is required. It is not common to perform more than two wash steps (see Note 6).

10. Retain the lid of the Falcon tube, dry it at 105°C and return it to the desiccator.

11. Place aluminium foil over the top of the tube, piercing it a few times to allow evaporation.

12. Place the tube upright in a drying oven at 105°C until a constant weight is achieved, typically 12–36 h.

13. Place the covered tube in a desiccator, with the retained lid, and allow it to reach room temperature.

14. Remove the foil, replace the lid, and weigh within 15 min of removal from the desiccator.

15. Subtract the weight of the dry, empty tube from the total to give the dry cell mass for the starting volume measured.

16. Scale the weight of cells to the preferred specific volume basis.

3.2. Total Cell Counts and Viability Counts

A total direct count of individual cells is the preferred method for determining cell number. In combination with a vitality stain, such as trypan blue, the number of viable cells can be determined. Alternatively, the number of colony forming units (CFU) can be determined by spreading a known sample volume over an agar plate and incubating the plate to develop the colonies. However, this method does not distinguish between colonies formed by a single cell or a clump of cells, and the growth conditions during incubation of the plate may not be representative of the monitored bioprocess. Therefore, direct counts are the more reliable reference cell count method.

1. Ensure the haemocytometer and cover-slip are clean and grease-free (see Note 9).

2. Affix the cover-slip to the haemocytometer (see Note 10).

3. Mix equal volumes of 0.4% trypan blue stain and homogeneous cell suspension.

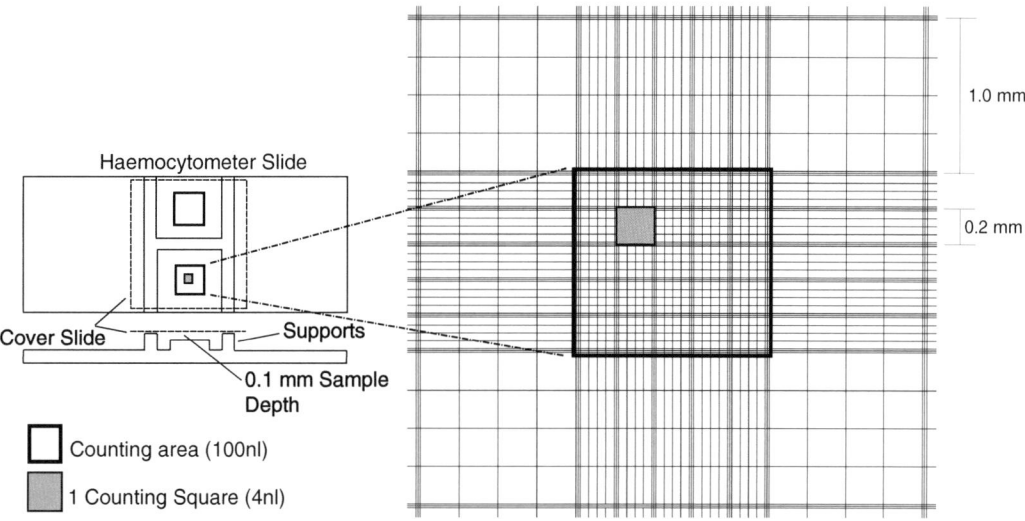

Fig. 1. Haemocytometer slide with detailed counting chamber. The cover slide ensures the sample depth is 0.1 mm over the counting grid, and gives a defined volume to the counting area.

4. Pipette the trypan blue/cell mixture at the edge of the cover-slip and allow it to run under the cover slip by capillary action. Approximately 10 µL is sufficient, but add enough to ensure that some leaks out of the cover glass. The volume required should be the same every time the haemocytometer is used.

5. Place the haemocytometer grid under the microscope and focus the objective onto a single counting square, as illustrated in Fig. 1.

6. Trypan blue is a vitality stain, meaning it is excluded from live cells. Live cells appear colourless, whereas dead cells take up and retain the blue stain, appearing blue.

7. If a total count is required, count all cells, stained and unstained. When viability is required, tally the viable and dead cells individually and calculate the percentage. If only viable cells are to be counted, count only the unstained cells.

8. To ensure accuracy and consistency when counting cells on the boundaries, count the cells within the large square and those crossing the middle line on two out of the four sides, as shown in Fig. 2 (see Note 11).

9. The number of individual squares to be counted depends on the concentration of cells. It is preferable to count at least 70 cells, but the larger the number of cells counted the more reliable the result.

10. If the cell count is high, further dilution of the sample may be needed, particularly if the cells are crowded and difficult to count without confusion. Any subsequent dilution must be factored into the calculations.

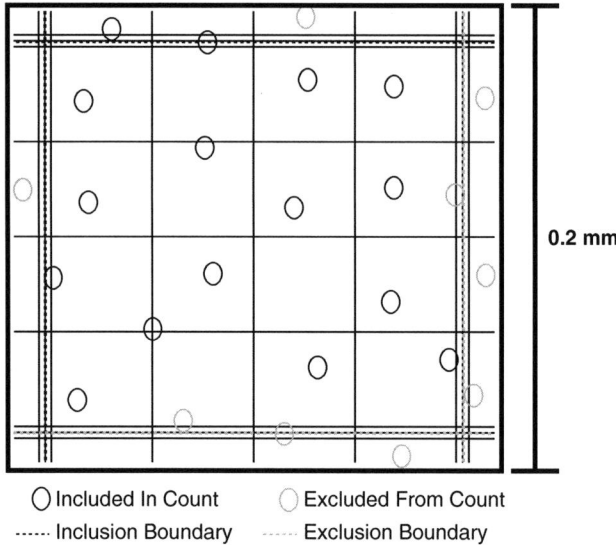

0.2 mm

◯ Included In Count   ◯ Excluded From Count
····· Inclusion Boundary   ····· Exclusion Boundary

Fig. 2. Expanded view of one counting square. Consistency in boundary cell counting is achieved by counting cells that cross the middle boundary line on two out of four sides.

11. Counting one shaded square 0.2 mm in width, illustrated in Fig. 1 as the counting area, gives the number of cells in 100 nL (10^{-4} mL) of diluted sample. Multiply the cell count per red square by 10^4, then by the dilution factor (which will be 2, if a 1:1 dilution with trypan blue is made) to get the number of cells per mL.

4. Notes

1. Unless stated otherwise, "water" refers to water that has a resistivity of less than 18.2 MΩ-cm and a total organic content of less than five parts per billion.

2. Centrifugation is much cheaper than filtration, but the required sample volume is much larger. A large cell pellet reduces error, but the actual volume required is primarily dictated by the concentration of cells in the sample. When centrifugation is used, care must be taken to ensure that cells are not withdrawn when decanting supernatant, such as during medium removal and cell washing. Centrifugation is not as rapid in arresting cellular activity as filtration. Although it is argued that such factors are within the standard error of gravimetric techniques, continued cellular activity can lead to either an increase (continued growth) or a decrease (mobilization of storage materials) in apparent cell mass. The arguments for cell washing are twofold. Cell weight could be overestimated when soluble components remain within and on the wetted surface of the cells but equally there

is an error associated with the leaching of intracellular components by washing. Cell washing is often considered a matter of preference, but consistency is important.

3. Filtration can be carried out rapidly, provided filters do not clog and arrest cellular activity more quickly than centrifugation. If only a small amount of cellular material is available, filtration is preferable because cells can be dried on the filter without risk of cell loss. The arguments for cell washing are precisely as stated above.

4. Cellulose filters can lose their mechanical strength at drying temperatures of 105°C, use nylon or polyvinylidene fluoride (PVDF) filters for robustness.

5. When the syringe is too large, it will contain compressible air, which makes force filtering more difficult.

6. Cell washing is, to some extent, a matter of preference. Whilst media components will remain in unwashed samples, soluble components may leach out of the cell during washing. Of course, if results from both protocols were compared, the errors become compounded, so consistency is a necessity.

7. Drying is most commonly carried out in an oven at 105°C, a few degrees higher than the boiling point of the solvent content (water). Freeze-drying is arguably a better option, as products of heat degradation are avoided, but freeze-dryers are expensive and less commonplace. One worker (1) reports a useful means of quick drying by using a microwave oven.

8. Use the smallest/narrowest Falcon tube possible for the volume to be handled. The smaller the interface between the cell pellet and the supernatant to be decanted, the less likely the pellet is to be disturbed.

9. Alcohol may be used to clean the haemocytometer. Lens-cleaning tissue should be used to wipe away any excess. After use, the haemocytometer should be rinsed with distilled water to remove any dye and sample, and allowed to dry. Haemocytometers should not require autoclaving, and doing so may warp the grid.

10. When the cover slide is fixed, the reflection of light between the two surfaces can be seen as Newton's rings, a series of rainbow-coloured concentric circles.

11. Usual practice is to count cells overlapping the top and left-hand lines, and to exclude those overlapping the bottom or right-hand lines (Fig. 2). Keeping to this system when counting multiple grids has the advantage of eliminating redundant counting when adjacent regions are counted.

Reference

1. Haack MB, Eliasson A, Olsson L (2004) On-line cell mass monitoring of *Saccharomyces cerevisiae* cultivations by multi-wavelength fluorescence. J Biotechnol 114:199–208

Chapter 14

Online Monitoring of Biomass Accumulation in Recombinant Yeast Cultures

Shane M. Palmer and Edmund R.S. Kunji

Abstract

Methods of biomass monitoring have increasingly been developed towards online, *in situ* techniques in order to advance process analysis and control. Off-line, *ex situ* methods, such as dry cell mass determination and direct cell counts, remain the reference for determining cell mass and number, respectively, but this type of analysis is time consuming. Absorbance measurement, which is used extensively as an off-line, *ex situ*, or online, *in situ* technique, is fast and straightforward, as the absorbance can be correlated to cell mass and number by a simple calibration. The downside is that absorbance measurements provide no estimation of viability and *in situ* applications can suffer from interference, such as aeration. Impedance spectroscopy is widely available and provides a quick measure of viable cell concentration, but does not give an estimation of total cell concentration and, hence, potential product. Sensitivity of impedance spectroscopy remains an issue at low cell concentration, and interference during *in situ* measurements is also a significant factor. In this chapter, a range of protocols is presented for online determination of biomass yields of recombinant yeast cultures.

Key words: Biomass, Accumulation, Monitoring, Yeast, Off-line, Online, *in situ*, *Ex situ*

1. Introduction

Decades of effort have yielded a spectrum of methods for biomass monitoring, but a diverse approach means a standard unit of biomass remains elusive. Off-line, at-line, and online systems have been developed, and both *in situ* and *ex situ* monitoring methods are available. For online *in situ* measurements, sensors have been developed that withstand wide-ranging process conditions, including repeated cleaning-in-place (CIP) and sterilization-in-place (SIP). Other desirable features for online sensors are their ease of operation and calibration, and a good operating range with minimal sensitivity to operating conditions and interference (1). Absorbance (transmittance) measurements fulfill the criteria of noninvasive,

Roslyn M. Bill (ed.), *Recombinant Protein Production in Yeast: Methods and Protocols*, Methods in Molecular Biology, vol. 866, DOI 10.1007/978-1-61779-770-5_14, © Springer Science+Business Media, LLC 2012

nondestructive, and continuous real-time measurement of biomass (2, 3). This method is popular because it is fast and easy to perform, like its off-line, *ex situ* equivalent (1). Additionally, the relationship of absorbance to cell counts and dry weight measurements has been shown to be acceptable for individual strains growing under defined conditions (4, 5). However, it should be noted that individual calibrations are required for any variation within a set of conditions. In particular, any change in condition that affects the osmotic environment of a culture will alter its turbidity for a given cell count or total mass, since fluid content influences directly cell size and consequently the amount of light scattering (2).

Impedance spectroscopy is another online method with significant popularity in yeast biomass monitoring. Appropriately reviewed since its first biological applications five decades ago (6–9), this technique offers an online viable cell mass representation, because only cells with a polarizable membrane are measured. The amount of polarization is measured as capacitance at a frequency where approximately half of the cells are polarized. Cells are fully polarized with a frequency below ~0.1 MHz. Since changes in the medium profile and other environmental factors affect overall polarization levels, a second reference frequency, often larger than 10 MHz, is used. At high frequency little cell polarization is achieved, so this polarization level can be used to adjust the apparent cell polarization caused by external factors.

In this chapter, a standard off-line procedure for measuring absorbance is presented followed by descriptions of online methods for absorbance and impedance (capacitance) monitoring.

2. Materials

2.1. Absorbance (Off-Line, Ex Situ)

1. Representative cell sample.
2. Reference sample.
3. Lidded sample storage tubes, such as Falcon tubes (Sarstedt, from VWR).
4. Cuvettes.
5. Spectrophotometer, such as WPA UV1101 Biotech Photometer (Jencons).
6. Syringe (Becton Dickinson, from VWR).
7. 25 mm syringe filter holder (Sartorius).
8. Filter paper (Whatmann GF/C, 25 mm, from VWR).
9. Vortex mixer (Scientific Industries, from Jencons).
10. Gilson pipettes (Anachem).

2.2. Absorbance (Online, In Situ)

1. Optical probe such as Hellma 661.760 process probe (Hellma).
2. Fiber optic cables with SMA905 connection (Ocean Optics).
3. Spectrophotometer such as USB2000 (Ocean Optics).
4. Light source such as LS-1 (Ocean Optics).
5. Software such as OOIBase32 (Ocean Optics).
6. Data Acquisition PC with USB port, Windows 2000 onwards.

2.3. Impedance Spectroscopy (Capacitance Method)

1. Capacitance monitor such as Model 200 (Aber Instruments).
2. Head amplifier and sensor cable (Aber Instruments).
3. Capacitance probe (Aber Instruments).
4. Probe Simulator (Aber Instruments).
5. Software such as Workbench 200 (Aber Instruments).
6. Data Acquisition PC with USB port, Windows 2000 onwards.

3. Methods

3.1. Absorbance (Off-Line, Ex Situ)

When light passes through matter across a path of a given length, it is scattered from its original path. The amount of scattering is dependent on many variables and not just those associated with cell population, such as concentration, size, and shape. The refractive index relative to the medium and the wavelength of incident light can also affect light scattering.

The Beer–Lambert law Eq. 1 states that there is a logarithmic dependence between the ratio of measured and incident light (transmittance) and absorbance. Absorbance Eq. 2 is the product of the absorptivity (ε) of a given concentration (c) of absorbing species (cells) over a given length (l), wherein that light is absorbed.

$$T = \frac{I}{I_0} = 10^{-\varepsilon l c} \tag{1}$$

$$A = \varepsilon c l \tag{2}$$

Transmittance, the raw measure of absorbance, is often presented as the percentage of incident light, which can be expressed as Eq. 3.

$$\%T = 100T \tag{3}$$

When Eq. 2 is substituted into the right hand term in Eq. 1, the mathematical term for absorbance can be simplified as Eq. 4.

$$A = \log_{10}\left(\frac{I_0}{I}\right) \tag{4}$$

Table 1
Absorbance at different transmittance values

Absorbance	Transmittance (I/I_0)	Percent transmittance (%T)
0	1.0	100
0.25	0.56	56
0.50	0.32	32
0.75	0.18	18
1.0	0.1	10
2.0	0.01	1
3.0	0.001	0.1

Substituting Eq. 1 in Eq. 4 produces Eq. 5, which can also be expressed as Eq. 6 by substitution of Eq. 3 into Eq. 5.

$$A = \log_{10}\left(\frac{1}{T}\right) \tag{5}$$

$$A = \log_{10}\left(\frac{100}{\%T}\right) \tag{6}$$

When the log term in Eq. 6 is separated Eq. 7 follows, which can be simplified further to give Eq. 8.

$$A = \log_{10}100 - \log_{10}\%T \tag{7}$$

$$A = 2 - \log_{10}\%T \tag{8}$$

Equation 8 is useful for calculating absorbance from transmittance and was used to generate the data in Table 1. Although some modern instruments have improved range, most commercially available nonlaser-based instruments are not reliable below 10% transmission (above 1 Absorbance Unit). Regardless of a stated instrument range, it is advisable to check the linear range for any sample, as it is a requirement for the application of the Beer–Lambert law. Therefore, samples should always be diluted to bring the absorbance into the linear range.

1. These instructions assume use of a single beam spectrophotometer with a single sample holder, but they are easily adaptable to double beam instruments, where the reference and sample can be read simultaneously, and to spectrometers with multiple sample stations.

2. Always use cuvettes of equivalent specification for a data set and ensure that the faces through which the light path will travel are free from scratches and blemishes.

3. Prepare the reference sample first, which should be a relevant cell-free representation of the measured sample.

4. Load one cuvette with the reference sample (see Notes 1 and 2).

5. Place the reference sample cuvette in the instrument, ensuring the measuring faces of the cuvette are orientated in the light beam. Make sure the cuvette is fully inserted in the cuvette holder.

6. Set the instrument to the required wavelength, e.g., 600 nm.

7. Press the reference button and wait for the instrument to read "zero."

8. It is good practice to "test" the reference sample by pressing the "test" button after zeroing to check that the instrument records a zero test value. If a zero reading for the reference sample is returned, the sample can now be measured.

9. Ensure the culture sample is homogenous by using vortexed samples.

10. Dilute the sample appropriately by using accurate pipettes. In general, the measured absorbance of the diluted samples should preferably read less than 0.6 to ensure accuracy within the linear range of the instrument.

11. Load three cuvettes for triplicate sample readings (see Notes 1 and 2).

12. Load each sample cuvette into the machine in turn and press the "test" button for each to get their individual absorbance. Record the readings and calculate the mean value. If samples have been diluted, multiply the mean value by the dilution factor.

13. When the measured sample absorbance is above 0.8, prepare samples with a higher dilution factor and repeat the measurements.

3.2. Absorbance (Online, In Situ)

Advances in different technologies have led to an increased use of *in situ* optical sensors. Fiber optic sensors and cables allow measurements to be transferred from source to the instrument and the development of charge-coupled devices (CCDs) and photodiode arrays, which replace expensive photomultipliers in conventional spectrometers, have made the technology more affordable and have extended the useful wavelength range.

The principles of online, *in situ* measurement are exactly the same as those in Subheading 3.1, but the online interface (software) requires some additional input to configure the continuous data acquisition. Other adjustments, such as intensity of the light source, are also required for specific systems.

These instructions describe the location of software functions from the main menu options. Many of the key functions can be configured in toolbars, for easier access, once the toolbar symbols are familiar. Although here we describe a USB2000 Ocean Optics device, most settings will apply to other systems.

1. Always ensure the measuring window of the *in situ* optical sensor is clean before assembling the closed system for sterilization. Use ethanol and lens cleaning tissue to clean any contamination (see Notes 3–5).

2. Check any sealing o-rings, such as port seals, before inserting probes into the closed system, replace as necessary.

3. Check all port fittings/adapters on the probe are secure.

4. Once the system is prepared and system parameters, such as temperature and pH, are stabilizing, it is possible to begin configuration of the data acquisition settings, prior to referencing.

5. Under the "Time Acquisition" menu function, select "Configure." In the submenu that opens, "Configure Acquisition" and "Configure Time Channels" options are given.

6. Choose "Configure Acquisition" to open a dialog box for the time acquisition processing and perform the following steps:

   (a) Tick the checkbox "Stream Data to Disk" to set the time acquisition parameters of this channel.

   (b) Tick the checkbox "Show Values in Status Bar" to display the most current time acquisition values in the status bar during the time acquisition.

   (c) Set the frequency with which data should be streamed to disk by entering a number of acquisitions in the data box (see Note 6).

   (d) In the "Stream and Autosave Filename" data box enter a location and filename for the streamed data to be saved.

   (e) Tick the "Save Full Spectrum with Each Acquisition" checkbox, if you wish to save complete spectral data (full waveband) for every enabled channel with every acquisition. Channels are enabled in the "Configure Time Channels" dialog boxw described in instructions step 9–10.

   (f) Set an initial delay, if required. This delays the start of data acquisition by the specified time, once the data acquisition is started.

   (g) Specify the frequency at which readings are to be acquired, which is the time elapsed between readings.

   (h) Data can be collected for a specified length of time or continually until the acquisition is manually stopped. To continue an acquisition until manual intervention, check the "Continue Until Manually Stopped" box. If the box is not checked, enter a suitable duration for the acquisition (see Note 7).

(i) Close the "Configure Acquisition" dialog box.

7. Under the "Time Acquisition" menu function, select "Configure." In the submenu that opens "Configure Acquisition" and "Configure Time Channels" options are given.

8. Choose "Configure Time Channels" to open a dialog box for the channel configuration.

9. The "Configure Time Channels" dialog box has eight tabs. Six different channel tabs, A–F, and two combination tabs. Select each channel to be configured in turn and set any combination functions to be configured after the individual channels are configured.

10. Each channel can be configured as followed:

    (a) Tick the "Enabled" checkbox to enable the current channel, allowing settings for the time acquisition parameters to be made.

    (b) Tick the "Plotted" checkbox if you wish to see a real-time graph plotted in the spectral window during acquisition.

    (c) Select the spectrometer channel to use for the acquisition.

    (d) Choose the wavelength (nm) at which data will be acquired for this channel.

    (e) Set the bandwidth (pixels). This defines the number of pixels on the CCD detector, around the analyzed wavelength, that the system will average.

    (f) Set a multiplicative factor (see Note 8) and additive factor (offset) if required. The equation applied before the software stores or plots the data is:

$$\text{Result} = (\text{Factor} \times \text{Reading}) + \text{Offset}$$

11. Configure other channels, as required, by selecting and enabling individual channels, B–F, from the tabs in the dialog box and repeating step 10.

12. If the combination of data from any two channels (A–F) is of interest, "Combination 1" and "Combination 2" allow a combination of the output values from any two channels to be made. The value of the two channels can be combined through addition, subtraction, multiplication, or division.

    (a) Tick the "Enabled" checkbox to enable the time acquisition combination to be calculated.

    (b) Tick the "Plotted" checkbox if you wish to see real-time combination plots in the spectral window during acquisition.

(c) Use the two channel dropdown menus to select the first and second channel to form the combination.

(d) Use the middle dropdown menu to select the mathematical function to be performed.

(e) In "Combination 2" the output of a configured "Combination 1" channel may also be used as either the first or second channel.

(f) A multiplication factor and offset can also be added, as with standard channel configuration.

(g) Close the dialog box.

13. The system to be monitored should now be fully prepared, such that the medium to be measured and the process conditions are a stable representation of those to be monitored. For example, in fermentation monitoring, any heat sensitive components that cannot be part of the SIP cycle should be added after sterilization, but before referencing.

14. To begin referencing the instrument, select "Spectrum" from the main menu list and select "Scope Mode."

15. In "Scope Mode," the OOIBase32 software displays a graph of intensity counts over the entire measurable waveband. The maximum intensity count, i.e., the peak of the curve should be set to around 3,500 counts. Small adjustments of intensity count can be achieved by varying the integration time (see Note 9). Alter the integration time by selecting "Configure Data Acquisition" under the "Spectrum" main menu. Larger adjustments can be made by attenuation of the input light, at source (see Note 10).

16. The OOIBase32 software offers two variables for data smoothing. Both are located in the "Configure Data Acquisition" dialog box, along with integration time. Entering a value in the "Spectra to Average" data box sets the number of discrete data acquisitions that the device driver accumulates, and averages, before OOIBase32 receives the spectrum. The signal-to-noise ratio improves by the square root of the number of scans averaged. The second tool, "Boxcar Smoothing Width," averages across spectral data by obtaining a mean value from the set number of adjacent detector elements. For example, if a value of 5 is entered, an average value using 5 data points to the left and right is taken. The signal-to-noise ratio also improves by the square root of the number of data points averaged but if the Boxcar value is too high a loss of spectral resolution results.

17. Store a dark spectrum (for all enabled channels, unless otherwise configured in "Misc. Settings") by switching off the light source and selecting "Store Dark" in the "Spectrum" menu (see Note 11).

18. Turn the light source back on (see Note 12).

19. Select "Store Reference" in the "Spectrum" menu to store a reference spectrum (see Notes 13 and 14).

20. In the "Spectrum" menu, select "Absorbance Mode." A graph of wavelength vs. absorbance is displayed. If the reference is stable and there is limited interference to cause fluctuations, the absorbance trace across the median waveband range (500–900 nm) should be approximately zero. If there is fluctuation because of interference, the average reading should be zero. If the average is obviously offset from zero, return to step 14, otherwise proceed.

21. Before data acquisition can be started, the time acquisition mode must be activated. In the "Time Acquisition" menu, select "Activate Time Acquisition." This does not begin the actual data acquisition, but activates the system for data acquisition to be enabled.

22. Begin the time acquisition by reselecting the "Time Acquisition" menu and selecting "Start." Data acquisition is now started. If the "Show Values in Status Bar" box was checked in step 4, the absorbance value for the first acquisition should be displayed in the status bar.

23. If the "Continue Until Manually Stopped" box was checked in the time acquisition configuration box during step 6, time acquisition will continue until stopped. If a suitable duration was set for the acquisition, time acquisition will continue for the specified time.

24. To manually stop a data acquisition, select "Stop" in the "Time Acquisition" menu. If the "Stream Data to Disk" box was checked, OOIBase32 immediately streams any outstanding data to disk.

25. All data will be stored in the location specified in point (d) of step 6. The files can be resaved as Excel files for data handling.

3.3. Impedance Spectroscopy (Capacitance Method)

This method measures the capacitance of the culture and assumes the use of an Aber Instruments Model 200 Biomass Monitor (BM200), running through Workbench 200 Software on a PC. The method is presented in two sections. First, the basic method of starting an experiment is covered. Thereafter, procedures for probe maintenance and recalibration are outlined.

3.3.1. Basic Experimental

1. Ensure the capacitance sensor is in good condition before assembling it in any closed system. Use the probe cleaning and maintenance procedures in Subheading 3.3.2 as guidance.

2. Carry out sterilization protocol and add any heat sensitive components, such that the reference sample is fully representative. Allow the system to stabilize at operating conditions.

3. Ensure that the BM200 is connected to the computer and powered up. Start the Workbench 200 software (see Note 15).

4. Log in to view the initial run dialog box with all options available.

5. Within the run dialog box, select an operating "Mode" in the dropdown menu. For yeast growth, "Microbial" mode is recommended (see Note 16). A custom mode is also available in the menu. All settings (see Note 16) can be customized by clicking "Custom Mode Settings." Settings can also be retrieved from a previous log file.

6. To retrieve settings from a previous log file, select "Get Settings from a Previous Log File" and select the desired file. Select "Save Settings to Instrument" to save and return to the run dialog box.

7. Select "Zero" to tare the instrument. Monitor the baseline reading for sufficient time to ensure stability (see Note 17).

8. If the baseline has settled away from zero, rezero the instrument. Data logging can now begin.

9. Select "Log Data" on the main menu to open the log data dialog box.

10. Choose an output file location and filename for the data. A title for the experiment must also be entered to activate the "Log Data" button. Use the comments box to record further detail. Personnel and date information are automatically logged by the BM200, and default to your windows login name and the current date, respectively. Both can be edited.

11. Set the reading frequency via the dropdown box. Intervals of 10 s, 30 s, or 1 min are possible.

12. If a calibration factor has been determined, see Subheading 3.3.3, the "Custom Units Instead of Capacitance" box can be checked to allow entry of an alternative measurement unit (e.g. cells/mL) and its multiplication factor. If you use this function, the software will automatically use the multiplication factor to convert capacitance biomass readings (pF/cm) to the chosen unit.

13. Once a filename and title have been entered in the Log Data dialog box, the "Start Logging" button is activated. Press "Start Logging" to begin data logging.

14. The "Data Capture in Progress" dialog box now appears, and a time course of capacitance and conductivity are plotted on-screen. For every parameter displayed, there are sets of buttons on the left hand side that allow adjustment of scale and zoom level of the data logged.

15. Select "Reset to Watching Current Graph" at any time to reset the scales and view the current graph.

16. To stop the data logging, select "Stop Logging." The software requires confirmation that you intended to stop the logging process. Confirm termination of the data logging to view the "Log Data from Instrument" dialog box.

17. In the "Log Data from Instrument" dialog box, there is an option to review the graph, or "Export to CSV" which allows direct export into Microsoft Excel (see Note 18).

18. The "View Data" button in the main dialog box opens up a "Log Data from Instrument" dialog box at any time. Previous data can be viewed and exported at any other time, via this route.

3.3.2. Probe and Instrument Maintenance

Regular inspection of the probe should be carried out to ensure that it is mechanically sound, including o-rings. Regularly check the electrodes for fouling, pitting, and erosion. Ensure that the probe plug and head amplifier are clean and dry. Routine probe cleaning can be carried out using ethanol. "Clean pulse" electrical cleaning is outlined below. Probe testing equipment, for full electrical testing, is available from Aber Instruments, with full instruction. The BM200 instrument can be tested using a probe simulator.

Fouling of capacitance probes can be an issue, and is dealt with through electronic (10 V) cleaning pulses, applied through the electrode, which generates gas bubbles by electrolysis. The idea is that the gas bubbles generated lift off any adherent material (10). This procedure is never carried out in the middle of an experiment, and can make readings unstable for up to 10 min.

1. Prepare a solution of sodium or potassium chloride in a beaker at the highest concentration possible.

2. With the head amplifier attached submerge the probe into the solution at least 20 mm beyond the electrodes.

3. In the main menu, select "Clean Pulse" and leave the probe to stand for at least 10 min.

4. Remove the probe and rinse in water (see Note 19).

5. To test the cleaned probe using a probe simulator, attach the head amplifier to the probe simulator. The BM200 automatically switches itself to "Test Mode."

6. Set the capacitance and conductance switches to 0 pF/cm and 0 mS/cm, respectively.

7. Zero the instrument in the main menu screen to remove the background capacitance measurement of the probe simulator.

8. Check the BM200 is in agreement with the parameters in Table 2 for the different simulator settings.

9. Remove the probe simulator from the head amplifier. The BM200 automatically returns to the previous mode.

Table 2
Expected reading from the BM200 with various probe simulator settings

Probe simulator: capacitance switch setting (pF/cm)	Probe simulator: conductance switch setting (mS/cm)	BM200 capacitance reading (pF/cm)	BM200 conductance reading (mS/cm)
0	0	0.0 ± 0.5	0.0 ± 0.1
100	0	100.0 ± 1.0	0.0 ± 0.1
0	2	0.0 ± 1.0	2.0 ± 0.1
100	2	Approx. 100[a]	Approx. 2[a]
0	20	Approx. 0[a]	20.0 ± 0.5
100	20	Approx. 100	Approx. 20

[a] Values are only really validated when the opposite switch is set to zero

3.3.3. Calibrating Capacitance (pF/cm) to Other Units of Biomass (E.g. cells/mL)

1. Carry out electrical cleaning of the probe, as described in Subheading 3.3.2.
2. Select a suitable operating mode for the cell type to be calibrated, i.e., microbial for yeast.
3. Place the probe in a cell-free medium, submerging the probe at least 20 mm beyond the electrodes.
4. Ensure the reading is stable which, for microbial culture settings, may take approximately 15 min.
5. Zero the instrument in the main menu, and check that the zero is stable.
6. Place the probe into a cell sample where the biomass concentration, in the units required, has been obtained by some other means. For example, calibrate capacitance to cells mL^{-1}, obtaining the true cell count via the method in Chapter 13.
7. Allow the reading to restabilize, and note the capacitance reading (pF/cm).
8. Use the capacitance reading, and the reference cell count as cells mL^{-1} to calculate a multiplication factor from Eq. 9.

$$\text{Multiplication Factor} = \frac{\text{Known Biomass}}{\text{Biomass Capacitance}} \qquad (9)$$

An individual multiplication factor is required for every strain and condition set used. The calculated factors can be entered into the Log Data screen, as described in step 12 of the basic experimental procedure.

3.3.4. Probe Calibration Factors: Uploading a Probe Table

Probes have a set of calibration factors that are specific to an individual probe. If a second probe is purchased for the BM200, it is necessary to upload a new set of calibration factors to the BM200, known as a probe table.

1. Save the new probe table file (supplied with the probe) onto your computer.
2. In the main menu, select "Setup" to enter the setup dialog box.
3. Click on the "Upload New Probe Table" button.
4. Select the file saved in step 1.

4. Notes

1. Load the sample slowly and evenly to minimize the introduction of bubbles. Take care not to dispense any sample onto the outer faces of the cuvette. Load the cuvettes close to capacity to ensure the light beam has passage through the sample and not air in the headspace above the sample.

2. Always handle cuvettes on their nonmeasuring faces. If fingerprints or other contaminants get onto the measuring faces, clean with a lens cleaning tissue and alcohol, and dry.

3. The quartz glass faces of the measuring window are highly polished and must not become scratched. Never use abrasive cloths or cleaners, consult the manufacturer if necessary.

4. Not all optical probes can be autoclaved. Although the internal sections exposed to the system environment are sealed, external interfaces are often unsealed.

5. Most optical probes come with a transmission test certificate that states the transmission loss through the probe at the time of manufacture. The transmission of a probe can be tested by comparing transmission traces with and without the probe in the optical loop.

6. Simply the number of discrete time acquisitions that the software performs before the software streams the data to disk. If reading frequencies are high, entering a higher number, reducing the number of times the software has to write to disk, can enhance the performance of the time acquisition.

7. It is worth noting that other software interfaces only allow the user to specify a frequency of acquisitions and the number of acquisitions to be made. In such cases, ensure the multiple of the two factors gives sufficient time for the monitored process to reach completion. Always check the units of time used in the configuration.

8. The multiplicative factor is where the use of different path length sensors should be compensated for. The default path length used in the Beer–Lambert law (see Eqs. 1 and 2) is 10 mm. For example, a factor of 10 needs to be used with a 1 mm path length, which is used to extend the upper measuring limit.

9. Integration time (ms) specifies the analogue/digital (A/D) conversion frequency of the spectrometer, and is analogous to the shutter speed of a camera. The higher the integration time, the longer the detector "looks" at the incoming photons. In our laboratory, we use a light attenuator at the light source to give a peak *close* to 3,500 counts, and then adjust the integration time, usually within the range 20–30 ms, for fine-tuning.

10. The integration time cannot be changed once the correction and referencing process is started. The "intensity" of input light must remain equivalent for dark correction, referencing, and sample analysis.

11. The dark correction spectrum is stored in temporary memory. To save these files permanently to disk, select "File," "Save," "Dark," and enter a filename.

12. Turning the light source off to allow a dark spectrum to be stored gives the light source time to cool, but always allow the bulb time to warm up again before taking the reference. It is advisable to turn the light source on at least 30 min before the referencing sequence is required. Use "Scope Mode" to check stability of the light intensity (counts) if necessary.

13. The dark correction spectrum is stored in temporary memory. To save these files permanently to disk, select "File," "Save," "Reference," and enter a filename.

14. Dark and reference data can be stored automatically for all enabled channels by selecting "Store All Active Spectrometer Channels" in the dropdown menu for "When Storing Reference and Dark." This can be found by selecting, "Edit," "Settings," and clicking on the "Misc. Settings" tab in the settings dialog box.

15. The "Login" button is not enabled unless the computer is detected.

16. These modes are sets of parameters that optimize the BM200 settings for particular cell types; modes include bacterial, microbial, and cell culture. For yeast monitoring (microbial), the biomass is measured at two frequencies 580 KHz and 15.65 MHz. At the lower frequency, cells in the microbial size range are well polarized. At 15.65 MHz very little cell polarization is achieved and any polarization is mainly from medium contribution and background matter. In microbial mode, the high frequency reading is automatically deducted from the low frequency reading to compensate background matter.

For microbial culture, the level of filtering is high. Filtering is implemented via a recursive digital filter in the software. Pinned probes are still preferred over annular probes for microbial measurement, but are prone to electrochemical affect at low frequency; as such a polarization correction is applied for microbial mode.

17. It is a useful practice to log the baseline data and subsequently have a graph on-screen to check baseline stability. Select "Log Data" to begin data logging and enter a filename in the "Results File" box. Suggested practice is to save the baseline data logs with a parallel filename to the main data files collected. A title must also be entered to activate the "Start Logging" button. Track the baseline readings for long enough to satisfy stability. With high filtering, this may take up to 15 min. Once a stable baseline is satisfied, select "Stop Logging" and "Close" to return to the main menu.

18. A "filename.abl" file format is also stored as a binary format to comply with validation legislation.

19. Attaching the probe simulator can test the "Clean Pulse" electronic cleaning function of the BM200. During the clean pulse, both LEDs on the simulator should glow red for a couple of seconds, green for a couple of seconds, then go out. You should never get different color LEDs illuminated at any time.

References

1. Olsson L, Nielsen J (1997) On-line and *in situ* monitoring of biomass in submerged cultivations. Trends Biotechnol 15:517–522

2. Mallette MF (1969) Evaluation of growth by physical and chemical means. In: Norris JR, Ribbons DW (eds) Methods in microbiology. Academic, London, England and New York, NY, USA, pp 521–566

3. Junker BH, Reddy J, Gbewonyo K, Greasham R (1994) Online and *in situ* monitoring technology for cell-density measurement in microbial and animal-cell cultures. Bioprocess Eng 10:195–207

4. Singh A, Kuhad RC, Sahai V, Ghosh P (1994) Evaluation of biomass. Adv Biochem Eng Biotechnol 51:47–70

5. Clarke DJ, Blakecoleman BC, Carr RJG, Calder MR, Atkinson T (1986) Monitoring reactor biomass. Trends Biotechnol 4:173–178

6. Fehrenbach R, Comberbach M, Petre JO (1992) Online biomass monitoring by capacitance measurement. J Biotechnol 23:303–314

7. Kiviharju K, Salonen K, Moilanen U, Eerikainen T (2008) Biomass measurement online: the performance of *in situ* measurements and software sensors. J Ind Microbiol Biotechnol 35:657–65

8. Carvell JP, Dowd JE (2006) On-line measurements and control of viable cell density in cell culture manufacturing processes using radiofrequency impedance. Cytotechnol 50:35–48

9. Markx GH, Davey CL (1999) The dielectric properties of biological cells at radiofrequencies: applications in biotechnology. Enzyme Microbial Technol 25:161–171

10. Yardley YE, Kell DB, Barrett J, Davey CL (2000) On-line, real-time measurements of cellular biomass using dielectric spectroscopy. Biotechnol Genetic Eng Rev 17:3–35

Chapter 15

Optimising *Pichia pastoris* Induction

Zharain Bawa and Richard A.J. Darby

Abstract

A common method for inducing the production of recombinant proteins in *Pichia pastoris* is through the use of methanol. However, the by-products of methanol metabolism are toxic to yeast cells and therefore its addition to recombinant cultures must be controlled and monitored throughout the process in order to maximise recombinant protein yields. Described here are online and off-line methods to monitor and control methanol addition to bench-top-scale bioreactors.

Key words: Methanol, Optimisation, Mixed feeds, Yeast

1. Introduction

The use of methanol as the sole carbon source for *Pichia pastoris* was first reported by Ogata and colleagues (1), who coined the term methylotroph to describe this yeast. Phillips Petroleum Company subsequently developed methods and protocols for culturing *P. pastoris* to high cell densities in methanol (2) and it has since been commercialised by Invitrogen Corporation, becoming a popular host system for recombinant protein production due to its tightly regulated *AOX* promoter (3).

This chapter focuses on the importance of optimising methanol feed rates during the cultivation of *P. pastoris* in order to produce maximal yields of recombinant protein. To appreciate why it is important to do this, we offer an insight into the relevant metabolic pathways of *P. pastoris*.

1.1. Methanol Metabolism in *P. pastoris*

The methanol utilisation pathway for *P. pastoris* is complex. The key points are summarised in Fig. 1. *P. pastoris* first oxidises methanol to formaldehyde via a flavoprotein, alcohol oxidase, whose expression is strongly suppressed by carbon sources, such as glucose, glycerol,

Roslyn M. Bill (ed.), *Recombinant Protein Production in Yeast: Methods and Protocols*, Methods in Molecular Biology, vol. 866, DOI 10.1007/978-1-61779-770-5_15, © Springer Science+Business Media, LLC 2012

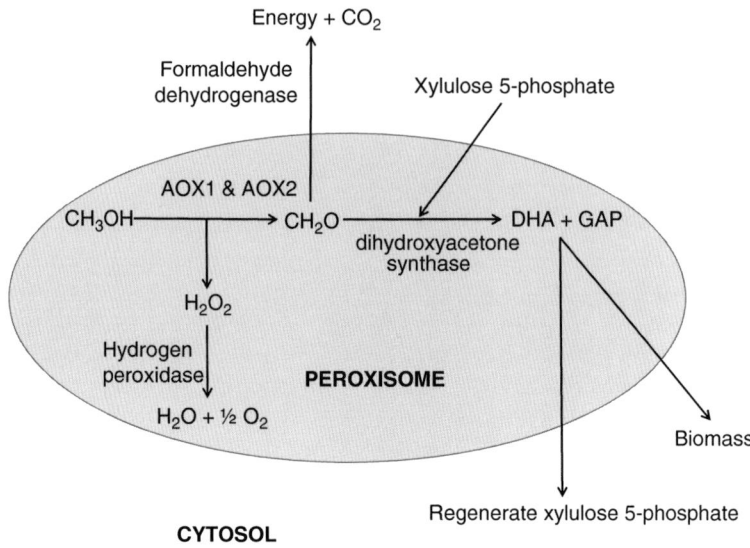

Fig. 1. A diagrammatic representation of the *Pichia pastoris* peroxisome outlining the methanol metabolism pathway.

and ethanol but which is induced in the presence of methanol and under conditions of carbon starvation. The reaction occurs within specialised compartments called peroxisomes and generates hydrogen peroxide. This metabolic by-product is toxic to the cell and it is subsequently catalysed by hydrogen peroxidase into water and oxygen.

The formaldehyde generated from the breakdown of methanol enters a cytosolic dissimilatory pathway to generate energy for the cells, as well as simultaneously activating an assimilatory pathway which results in an increase in cellular biomass. During this assimilatory pathway, residual formaldehyde reacts with xylulose-5-monophosphate and is converted to dihydroxyacetone and glyceraldehyde-3-phosphate by dihydroxyacetone synthase. These products leave the peroxisome and are used to regenerate xylulose 5-monophosphate and glyceraldehyde-3-phosphate (Fig. 1). It has been reported that both the alcohol oxidase and dihydroxyacetone synthase pathways are only detectable when *P. pastoris* is grown in the presence of methanol (3).

1.2. P. pastoris *AOX* Phenotypes

Wild-type *P. pastoris* has two endogenous copies of the *AOX* gene; *AOX1* expression accounts for more than 90% of the enzyme in the cell while *AOX2* expression constitutes less than 10%. Genetic engineering of these genes has generated three commonly used strains of *P. pastoris* that vary in their ability to metabolise methanol. The wild-type or methanol utilisation plus phenotype (Mut⁺) which has genes for both *AOX1* and *AOX2*; the methanol utilisation slow phenotype (Mutˢ) resulting from deletions to the *AOX1* gene and finally, strains with deletions to both the *AOX1* and *AOX2* genes yielding a methanol utilisation minus phenotype (Mut⁻) (4).

1.3. Optimising Methanol Induction

The tightly regulated expression of heterologous genes in *P. pastoris* is often achieved by using the promoter of *AOX1* in an expression vector, such as Invitrogen's pPICZ series. Cells are cultured in the presence of non-limiting amounts of glucose or glycerol to repress the expression of the gene followed by an induction phase in the presence of methanol (5). In practice, *P. pastoris* is often cultured in media with non-limiting amounts of glycerol to generate biomass, followed by an induction period with limiting amounts of methanol. Glycerol is the preferred carbon source as it permits de-repression and not expression of the *AOX* genes while glucose represses their expression even in the presence of methanol. Furthermore, it has been shown that there is a higher biomass yield when *P. pastoris* is grown in glycerol compared with either glucose or ethanol as carbon sources (6). Three distinct phases of the culture are termed the "batch phase" in the presence of glycerol, the "starvation phase" when the glycerol is depleted and the "fed-batch phase" when heterologous protein production is induced via the addition of limiting amounts of methanol (7).

Although the simple addition of methanol to a *P. pastoris* culture induces protein production, careful consideration must be given to the duration and rate of addition as methanol can be toxic to the cells if accumulation occurs within the culture (8). Conversely insufficient methanol will result in sub-optimal protein yields. It is therefore imperative to strike the optimal balance (9). The general rule-of-thumb is that the concentration of methanol within the culture should be maintained below 0.5% (v/v) to avoid cytotoxicity; however, this generic figure may not be optimal for all recombinant proteins and should be derived on a case-by-case basis (8). Furthermore, it is often common practice to induce expression in *P. pastoris* cultures solely using methanol; however, reports have demonstrated increased protein yields by inducing with a mixed feed of glycerol and methanol (10, 11) or sorbitol and methanol (12, 13). It is possible that both these strategies decrease the potential toxicity to the yeast cells caused by methanol overload, as well as permitting enhanced biomass generation in the induction phase compared with a solely methanol fed system.

1.4. Alternatives to Methanol Induction

Even though methanol induction is robust and tightly regulated, the health risks associated with methanol, i.e. its toxicity and volatility have led researchers to investigate alternative promoters that do not require its use (4). For example, constitutive expression has been observed when the glyceraldehyde 3-phosphate dehydrogenase (*GAP*) promoter is used (14, 15), while strong induction via the formaldehyde dehydrogenase (*FLD1*) promoter has also been reported in the presence of methylamine as well as methanol (16). As they are not widely used, they are not discussed further in this chapter.

2. Materials

- YPD agar: 1% yeast extract, 2% peptone, 2% glucose, 2% agar (*stable at 4°C up to 2 weeks*).
- BMGY medium: 1% yeast extract, 2% peptone, 100 mM potassium phosphate, pH 6, 1.34% yeast nitrogen base without amino acids, 4×10^{-5}% biotin, 1% glycerol (*stable at 4°C up to 2 weeks*).
- HPLC grade methanol.
- Glycerol (99.5% purity).
- *P. pastoris strain*: freshly plated.
- 60% D-sorbitol (*stable at 4°C up to 1 year*).
- Fermentation basal salts medium: 26.7 mL phosphoric acid (85%), 0.93 g calcium sulphate, 18.2 g potassium sulphate, 14.9 g magnesium sulphate·7H$_2$O, 4.13 g potassium hydroxide, 40 g glycerol, up to 1 L water (*autoclave* in situ *in the bioreactor and use within 48 h*).
- PTM$_1$ trace salts 1 L: 6.0 g copper sulphate·5H$_2$O, 0.08 g sodium iodide, 3.0 g manganese sulphate·H$_2$O, 0.2 g sodium molybdate·2H$_2$O, 0.02 g boric acid, 0.5 g cobalt chloride, 20.0 g zinc chloride, 65.0 g ferrous sulphate·7H$_2$O, 0.2 g biotin, 5.0 mL sulphuric acid, up to 1 L water (*filter sterilise, protect from light and store at 4°C for up to 1 year*).
- Antifoam A (Fluka Analytical).
- 250 mL baffled shake flasks.
- Sterile triple vented petri dishes (100 mm × 200 mm).
- Humidified incubator set at 30°C.
- Rotatory temperature controlled incubator.
- Spectrophotometer.
- Light microscope.
- 50 mL falcon tubes.
- 85% phosphoric acid.
- 28% ammonium hydroxide.
- 3 L jacketed glass bioreactor and controllers (Applikon Biotechnology).
- Off-gas analyser (Tandem).

- Methanol sensor system 2.1 (Raven Biotech Inc.).
- Gas chromatograph (Unicam 610 GLC Gas Chromatograph with ProGC software).

3. Methods

As discussed in Chapter 11, a statistical design may aid the success of producing high recombinant protein yields in yeast host systems by optimising culture components and conditions. In addition, further increases in recombinant protein yields can be achieved by optimising the methanol feed rates in the culture. The following methods outline protocols for optimising methanol induction of *AOX*-promoted genes using online and off-line methanol measurements as opposed to DO spike monitoring, which involves stopping the flow of the methanol and checking if the DO rises (spikes) as the culture decreases its metabolic rate. If this takes longer than 1 min, this suggests that methanol is accumulating (17). These methods can be applied to induction with methanol as the sole carbon source (18) or with a mixed-feed induction (12, 13).

3.1. Monitoring and Controlling Methanol Feed Rates On-Line

1. Plate out *P. pastoris* cells onto a YPD plate and incubate at 30°C in a humidified incubator for up to 5 days.

2. Aseptically pick a single *P. pastoris* colony from the agar plate into 50 mL BMGY medium in a 250 mL baffled shake flask and culture at 30°C with 220 rpm agitation for up to 48 h (see Note 1).

3. Prepare 1.5 L fermentation basal salts medium (see Note 2), transfer it into a 3 L jacketed glass bioreactor and autoclave.

4. Aseptically add 4 mL PTM_1 salts/L basal salts medium in the bioreactor (see Note 3).

5. Equilibrate the bioreactor to the run conditions (see Note 4).

6. Set up the methanol sensor probe and its respective controller, software and external pump. Perform a standard curve for methanol using 1 g/L as the low point standard and 5 g/L as the high point standard, according to the manufacturer's instructions.

7. Aseptically transfer sufficient pre-culture into the bioreactor to achieve a final OD_{600} of 1 (see Note 5).

8. Start the controlling software. This is the start of the batch phase and should be maintained for 20–24 h (see Note 6) to ensure that most of the glycerol in the basal salts medium has been consumed (60 g in 1.5 L).

9. The culture should now enter the glycerol fed-batch phase whereby a limiting amount of glycerol is pumped into the bioreactor for a period of 4 h (see Note 7).

10. A starvation period now occurs (see Note 8) during which the bioreactor conditions should be adjusted to the required induction phase conditions.

11. The methanol pump can now be switched on to start induction of recombinant protein production. If a 100% methanol stock is being used, the recommended rate of delivery should initially be between 1 and 2 mL/h.

12. It is important at this stage to monitor the methanol sensor readout and be prepared to decrease the flow rate of the methanol if the level reaches 5 g/L (see Note 9).

13. Continue the experiment for a given period of time (70–100 h), continuously monitoring the methanol level and maintaining it below 5% (see Note 10) by manually increasing or decreasing the flow rate as appropriate, or by maintaining a set-point value via a feedback loop.

3.2. Monitoring Methanol Off-Line

If equipment such as the Methanol sensor system 2.1 (Raven Biotech Inc.) is not available to monitor methanol levels online, the following steps can be taken to determine the level of methanol off-line.

1. At the end of the starvation phase, take a sample of the culture in a sterile sampling tube, centrifuge at $3,000 \times g$ for 3 min, recover 1 mL of the supernatant and store at $-20^\circ C$. Store the pellet at $-80^\circ C$. This supernatant will be the baseline sample before methanol induction and this sampling process should be repeated for all subsequent sampling periods.

2. The methanol pump can now be switched on to start induction of recombinant protein production with a recommended flow rate of 1–2 mL/h.

3. Take samples periodically (every 2–5 h) and store them as described above.

4. The methanol flow rate should be increased to 3 mL/h, 5 h post-induction and 6 mL/h, 12 h post-induction (see Note 11).

5. Once the process is complete (usually 72 h) the stored supernatants can be analysed for methanol via gas chromatography (see Note 12). A calibration curve with known concentrations of methanol should be prepared prior to analysing the samples. We recommend the following range; 0, 0.02, 0.05, 0.1, 0.2, 0.5, and 1% methanol diluted in basal salts medium containing PTM_1 salts to provide an accurate background.

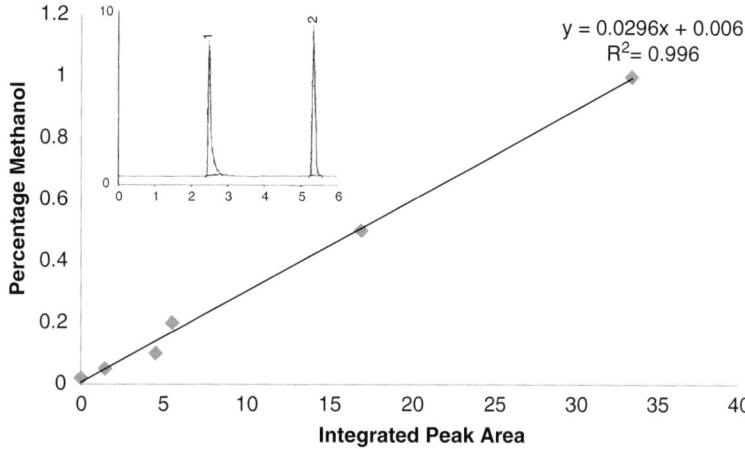

Fig. 2. A representative calibration curve for determining the concentration of methanol within the bioreactor at the time of sampling. *Inset* is a representative gas chromatograph for a dual injection of one of the methanol standards from which the integrated peak areas are calculated.

6. A calibration curve should be plotted of the gas chromatograph peak areas against the methanol concentration. Figure 2 gives an example of a calibration curve.

7. 1 μL supernatant should be injected in duplicate or triplicate and the average methanol concentration present in the samples derived from the calibration curve.

8. If any sample is outside the linear range of the calibration curve, dilute the sample appropriately and repeat.

4. Notes

1. The optical density of the culture should be determined at 600 nm and must be 15–30 before being used to inoculate the bioreactor.

2. We use Applikon Biotechnology bioreactors and associated controllers and the set-up may differ from other manufacturers. The medium should be prepared as stated and the pH recorded. Transfer the medium to the bioreactor and secure the head plate with all the autoclavable probes (pH probe, DO probe, optical density probe, methanol sensor, off-gas condenser) in situ. Calibrate the bioreactor pH probe to match the off-line pH value, according to the manufacturer's instructions. Autoclave the bioreactor vessel at 121°C, 20 min, 1.5 bar and allow to slow cool to prevent medium boil-off.

3. The PTM_1 salts should not be autoclaved. They should be prepared as stated and aseptically added to the bioreactor once it has cooled to room temperature and the medium adjusted to pH 5. A cloudy precipitate may form during this addition, which does not have adverse effects.

4. These conditions may have been established from earlier trials. However, if this is not the case, then we would recommend starting with conditions of 30°C, pH 5, >20% dissolved oxygen (DO).

5. The pre-culture should be checked for microbial contamination before inoculating the bioreactor. 10 μL of the pre-culture should be viewed using a light microscope with a 100× oil immersion objective. If there are any contaminating microbes present in the sample, the culture should be discarded and a fresh pre-culture prepared.

6. We have found that on average the batch phase last 20–24 h. The exact time can be established by monitoring the CO_2 level in the off-gas and the DO in the bioreactor as these will decrease and increase, respectively, as the glycerol is depleted. These levels should not be allowed to return to their baseline values.

7. Upon observing the profile changes in Note 6, the glycerol feed pump should be activated at a low flow rate and maintained by the system. This will permit sufficient glycerol to be added to the system to maintain growth without leading to excessive accumulation of the carbon source.

8. A period of 1–2 h without any further addition of glycerol is maintained allowing any accumulated glycerol from the previous phase to be consumed. Once again careful monitoring of the CO_2 level in the off-gas and the DO in the bioreactor are recommended. These levels should be allowed to begin decreasing and increasing, respectively, but not to return to their baseline values. Once these changes are observed the induction parameters should be set.

9. This can be achieved by manually adjusting the flow rate of the methanol pump or by setting up a feedback loop between the methanol sensor monitor and the methanol feed pump. Methanol levels above 5 g/L will result in cytotoxicity, an increase in methanol accumulation, and sub-optimal recombinant protein yields.

10. As well as monitoring the methanol level, a good indication of the health of the culture is the CO_2, DO, and optical density. As the yeast cells metabolise methanol, they produce CO_2, consume O_2, and generate biomass. If the level of CO_2 in the off-gas drops, the DO increases, or the optical density drops, these are markers for non-optimal conditions and methanol

addition should be stopped and the methanol sensor checked for accuracy by measuring a known concentration of methanol. Biomass should increase throughout the fermentation; however, growth is more rapid during the glycerol phases.

11. These flow rates are based on feeding with methanol as the sole carbon source at a process temperature of 30°C. Using a mixed feed stock or a different process temperature will require the flow rates to be derived on a case-by-case basis.

12. These data represent the level of methanol present at the sampling time and can be used to adjust the methanol flow rates in subsequent processes assuming that all the bioreactor conditions are maintained and the same yeast stock is used.

References

1. Ogata K, Nishikaw H, Ohsugi M (1969) A yeast capable of utilizing methanol. Agr Biol Chem 33:1519

2. Wegner GH (1990) Emerging applications of the methylotrophic yeasts. FEMS Microbiol Rev 87:279–283

3. Cereghino JL, Cregg JM (2000) Heterologous protein expression in the methylotrophic yeast *Pichia pastoris*. FEMS Microbiol Rev 24:45–66

4. Cos O, Ramon R, Montesinos JL, Valero F (2006) Operational strategies, monitoring and control of heterologous protein production in the methylotrophic yeast *Pichia pastoris* under different promoters: a review. Microb Cell Fact 5:17

5. Faber KN, Harder W, Ab G, Veenhuis M (1995) Review – methylotrophic yeasts as factories for the production of foreign proteins. Yeast 11:1331–1344

6. Inan M, Meagher MM (2001) Non-repressing carbon sources for alcohol oxidase (*AOX1*) promoter of *Pichia pastoris*. J Biosci Bioeng 92:585–589

7. Minning S, Serrano A, Ferrer P, Sola C, Schmid RD, Valero F (2001) Optimization of the high-level production of *Rhizopus oryzae* lipase in *Pichia pastoris*. J Biotechnol 86:59–70

8. Guarna MM, Lesnicki GJ, Tam BM, Robinson J, Radziminski CZ, Hasenwinkle D, Boraston A, Jervis E, MacGillivray RTA, Turner RFB, Kilburn DG (1997) On-line monitoring and control of methanol concentration in shake-flask cultures of *Pichia pastoris*. Biotechnol Bioeng 56:279–286

9. Thorpe ED, d'Aanjou MC, Daugulis AJ (1999) Sorbitol as a non-repressing carbon source for fed-batch fermentation of recombinant *Pichia pastoris*. Biotechnol Lett 21:669–672

10. d'Anjou MC, Daugulis AJ (2000) Mixed-feed exponential feeding for fed-batch culture of recombinant methylotrophic yeast. Biotechnol Lett 22:341–346

11. Jungo C, Schenk J, Pasquier M, Marison IW, von Stockar U (2007) A quantitative analysis of the benefits of mixed feeds of sorbitol and methanol for the production of recombinant avidin with *Pichia pastoris*. J Biotechnol 131:57–66

12. Ramon R, Ferrer P, Valero F (2007) Sorbitol co-feeding reduces metabolic burden caused by the overexpression of a *Rhizopus oryzae* lipase in *Pichia pastorism*. J Biotechnol 130:39–46

13. Holmes WJ, Darby RAJ, Wilks MDB, Smith R, Bill RM (2009) Developing a scalable model of recombinant protein yield from *Pichia pastoris*: the influence of culture conditions, biomass and induction regime. Microb Cell Fact 8:35

14. Waterham HR, Digan ME, Koutz PJ, Lair SV, Cregg JM (1997) Isolation of the *Pichia pastoris* glyceraldehyde-3-phosphate dehydrogenase gene and regulation and use of its promoter. Gene 186:37–44

15. Kim SJ, Lee JA, Kim YH, Song BK (2009) Optimization of the functional expression of *Coprinus cinereus* peroxidase in *Pichia pastoris* by varying the host and promoter. J Microbiol Biotechnol 19:966–971

16. Resina D, Cos O, Ferrer P, Valero F (2005) Developing high cell density fed-batch cultivation strategies for heterologous protein

production in *Pichia pastoris* using the nitrogen source-regulated *FLD1* promoter. Biotechnol Bioeng 91:760–767

17. *Pichia* fermentation process guidelines. http://tools.invitrogen.com/content/sfs/manuals/pichiaferm_prot.pdf

18. Nyblom M, Öberg F, Petersson-Lindkvist K, Findlay H, Wikström J, Karlsson A, Hansson O, Booth PJ, Bill RM, Neutze R, Hedfalk K (2007) Exceptional overproduction of a functional human membrane protein. Prottein Expr Purif 56:110–120

Chapter 16

Optimizing *Saccharomyces cerevisiae* Induction Regimes

David Drew and Hyun Kim

Abstract

Recombinant membrane protein yields can be optimized in *Saccharomyces cerevisiae* by adjusting the induction time and temperature and/or by the addition of chemical chaperones. Here we describe a protocol for assessing the importance of these parameters.

Key words: Membrane protein, Overproduction, *Saccharomyces cerevisiae*

Abbreviations

GAL   Galactokinase
GFP   Green fluorescent protein
GPCRs   G protein-coupled receptors
TEF   Translation elongation factor 1α

1. Introduction

Tuning membrane protein production is an empirical process. To facilitate optimization we tag all membrane proteins with green fluorescent protein (GFP). This strategy has previously made it feasible to designate the subcellular localization of approximately 75% of the *Saccharomyces cerevisiae* proteome (1). There are also many examples of specific localization studies in yeast using GFP fusions in combination with assays of protein function (2–4). These support the view that fusion with GFP does not usually perturb function.

Roslyn M. Bill (ed.), *Recombinant Protein Production in Yeast: Methods and Protocols*, Methods in Molecular Biology, vol. 866, DOI 10.1007/978-1-61779-770-5_16, © Springer Science+Business Media, LLC 2012

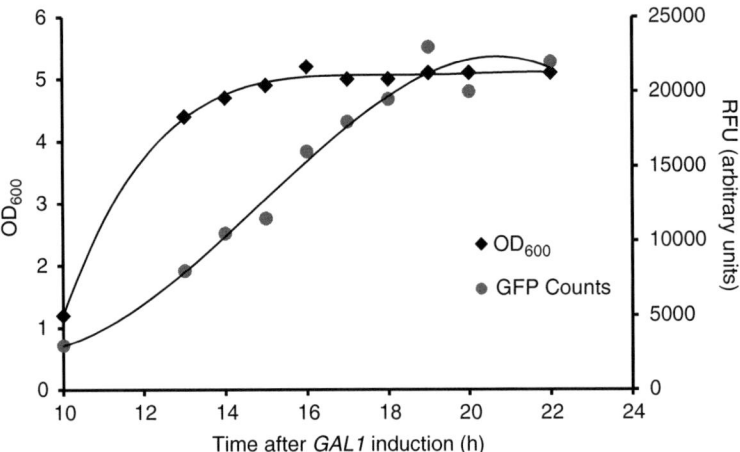

Fig. 1. Time course for the overexpression of a membrane protein–GFP fusion as monitored by GFP fluorescence (*circles*) and OD_{600} (*diamonds*) after induction with galactose in the FGY217 strain (9) under the control of the *GAL1* promoter (5).

We previously analyzed the recombinant yields of 20 yeast membrane protein–GFP fusions under the control of a constitutive *TEF2* and an inducible *GAL1* promoter (5). The majority of yeast membrane protein–GFP fusions exhibited higher yields under the control of the *GAL1* promoter compared with the *TEF* promoter (5). We were able to optimize induction times for optimal expression of membrane protein–GFP fusions in the former system using assays of fluorescence intensity. Most noticeably we found that the highest yields typically occur 12–16 h after the addition of galactose (Fig. 1).

Some studies have shown that addition of chemical chaperones such as dimethyl sulfoxide (DMSO), histidine, or glycerol, as well as lowering the cultivation temperature from 30°C to 20°C, can improve functional yields for a number of G protein-coupled receptors (GPCRs) in *Pichia pastoris* (6) as well as several other transporters in *S. cerevisiae* (7). We found that in most cases yields were enhanced by the addition of chemical chaperones (Fig. 2). We also compared induction times at both 30°C and 20°C and found that at least 36 h are required for growth at 20°C to achieve yields comparable with those at 30°C after 24 h. Using GFP fluorescence at different time points in the cultivation, both localization and monodispersity of the recombinant membrane protein–GFP fusions, could be assessed, information on which is a good indicator for the quality of the protein.

Recently, we tested an alternative induction scheme, whereby membrane protein production is induced in a nonselective rich medium such as yeast extract/peptone/dextrose (YPD). If protein production in a selective medium is nontoxic to the cells, i.e., if the culture can achieve a final OD_{600} greater than 7, then the plasmid

	A.t Ysl1	H.s Ctr1	H.s Sialin	Yea4p	Ctr3p	A.t Ysl2	Hut1p	Pho87p	M.m CMP-Sia-Tr	C.e Sqv7	MrH1p	Vrg4p	H.s Slc35 b1	H.s CMP-Sia-Tr	Azr1p	Isc1p	Kha1p	Hsp30p	Dur3p	Rft1p
■ Standard (mg/L)	0.0	0.1	0.1	0.4	0.4	0.6	0.7	1.0	1.0	1.2	1.4	1.7	1.7	1.7	2.1	2.2	3.1	3.3	3.3	3.5
☐ Glycerol (mg/L)	0.0	0.0	0.0	0.1	0.2	0.2	0.1	0.3	0.7	0.3	0.7	0.9	0.8	1.9	3.2	2.6	3.7	0.7	3.1	4.3
■ DMSO (mg/L)	0.0	0.1	0.2	0.6	0.1	0.8	0.9	1.2	1.2	1.4	1.2	1.3	1.7	2.1	4.1	3.2	3.1	2.8	4.1	4.2
☐ Histidine (mg/L)	0.0	0.1	0.1	0.2	0.4	0.6	0.9	1.2	1.0	1.3	1.4	2.0	1.7	1.9	2.5	3.5	3.1	3.0	4.0	3.2
☐ 20°C 36 hrs (mg/L)	0.3	0.1	0.1	0.5	0.2	0.4	0.4	1.0	0.7	0.7	1.9	1.8	0.3	1.3	2.4	1.8	3.0	2.7	2.7	3.0
☐ 20°C 24 hrs (mg/L)	0.1	0.0	0.0	0.2	0.5	0.3	0.2	0.4	0.4	1.1	0.1	0.8	0.1	0.8	3.7	1.0	1.7	1.3	1.6	3.1

Fig. 2. Mean yields (mg/L) from cultures induced under various conditions: standard growth medium lacking uracil; addition of glycerol (10% w/v); addition of DMSO (2.5% w/v); addition of histidine (0.04 mg/mL); growth at 20°C for 36 h or growth at 20°C for 24 h. For each membrane protein–GFP fusion, the best-yielding condition is shaded and is at least 0.2 mg/L (5).

is likely to be retained even in a nonselective medium. In these cases a higher biomass, in the region of $OD_{600} = 15$, is achievable while retaining the same yields per OD_{600} unit. Thus, it is possible to improve yields by more than a factor of three. Using this scheme we have produced the human glucose transporter, GLUT-1, at yields of more than 7 mg/L (8).

2. Materials

- Materials for yeast transformation and growth media are found in Chapter 4.
- DMSO (Sigma).
- Histidine (Sigma).
- Glycerol (Sigma).
- 20% galactose (w/v; Sigma).
- YSB (yeast suspension buffer): 50 mM Tris–HCl (pH 7.6), 5 mM EDTA, 10% glycerol, 1× complete protease inhibitor cocktail tablets (Roche).
- 96-well black optical-bottom plates (Nunc).
- SpectraMax M2e microplate reader (Molecular Devices).
- Confocal microscope (TCS SP2 upright confocal microscope; Leica).

3. Methods

1. Membrane protein–GFP fusions that are produced well under standard culture conditions are good candidates for further optimization. Thus, inoculate 10 mL medium lacking uracil with 2% glucose with an appropriate yeast colony (see Chapter 8) in 50-mL aerated capped tubes.

2. Dilute the overnight culture to an OD_{600} of 0.1–0.12 into six 50-mL aerated capped tubes, each containing 10 mL of medium lacking uracil with 0.1% glucose. Label duplicate tubes as follows: no chaperone; + DMSO; + histidine. Incubate at 30°C, 280 rpm.

3. Monitor the OD_{600} of the cultures. At $OD_{600}=0.6$ (after approximately 7 h) induce with galactose to a final concentration of 2% w/v. Add DMSO (2.5% w/v) or histidine (0.04 mg/mL) to the correspondingly labelled tubes (see Note 1).

4. 22 h after induction, measure the final OD_{600}, centrifuge the cells at $3,000 \times g$ for 5 min, remove the supernatant and resuspend the cell pellet in 200 μL YSB.

5. Transfer 200 μL cell suspension to a black Nunc 96-well optical-bottom plate (see Note 2).

6. Measure the GFP fluorescence emission at 512 nm following excitation at 488 nm in a microtitre plate spectrofluorometer. For plate readers with a bottom-read option, choose this setting. Estimate membrane protein yield (in mg/L) from the yeast whole-cell fluorescence reading by applying the methodology detailed in Table 1 in Chapter 8.

7. Repeat the culture condition that gave the highest yield (steps 1–6) at a post-induction (after addition of galactose) temperature of both 30°C and 20°C. For the latter, induce cells for 36 h (see Note 3).

8. After induction for 22 h at 30°C or 36 h at 20°C, harvest the cells as in step 4 and resuspend in medium lacking uracil with 50% glycerol.

9. Add 1 μL cell suspension to a microscope slide and place a cover slip on top. Focus on the sample using Köhler illumination at 10× magnification. Set the focal plane to 0.

10. Add a drop of lens oil and change to a higher magnification, oil-immersion lens. Turn off the bright field lamp. Turn on the blue light to check the total number of fluorescent cells. Turn off the blue light. Use the Argon laser emitting at 488 nm to image the cells.

11. Judge the optimal induction regime based on the conditions that give the highest yield of correctly targeted membrane protein–GFP fusion (see Note 4).

4. Notes

1. For initial optimization, we recommend the addition of DMSO or histidine as we find that most often these give an average yield improvement of 30%.

2. Because yeast cells settle to the bottom of the plate, proceed to the next step within 5 min of transfer to ensure accurate measurements.

3. Cells grow more slowly at 20°C, thus a longer incubation time is needed to increase biomass to the same levels as cultures grown at 30°C.

4. In addition to localization by confocal microscopy, the quality of the recombinant material can also be assessed by fluorescence-detection size exclusion chromatography (FSEC; Chapter 8). Functional assays should always be carried out where possible.

Acknowledgments

This work was supported by the Royal Society (United Kingdom) through a University Research Fellowship to DD and by a Basic Science Research Program grant through the National Research Foundation of Korea (NRF) funded by the Ministry of Education, Science and Technology (NRF0409-20100093) to HK.

References

1. Huh WK, Falvo JV, Gerke LC, Carroll AS, Howson RW, Weissman JS, O'Shea EK (2003) Global analysis of protein localization in budding yeast. Nature 425:686–691

2. Campbell SG, Ashe MP (2007) An approach to studying the localization and dynamics of eukaryotic translation factors in live yeast cells. Methods Enzymol 431:33–45

3. Greene LE, Park YN, Masison DC, Eisenberg E (2009) Application of GFP-labeling to study prions in yeast. Protein Pept Lett 16:635–641

4. Guo Y, Au WC, Shakoury-Elizeh M, Protchenko O, Basrai M, Prinz WA, Philpott CC (2010) Phosphatidylserine is involved in the ferrichrome-induced plasma membrane trafficking of Arn1 in *Saccharomyces cerevisiae*. J Biol Chem 285:39564–39573

5. Newstead S, Kim H, von Heijne G, Iwata S, Drew D (2007) High-throughput fluorescent-based optimization of eukaryotic membrane protein overexpression and purification in *Saccharomyces cerevisiae*. Proc Natl Acad Sci USA 104:13936–13941

6. André N, Cherouati N, Prual C, Steffan T, Zeder-Lutz G, Magnin T, Pattus F, Michel H, Wagner R, Reinhart C (2006) Enhancing functional production of G protein-coupled receptors in *Pichia pastoris* to levels required for structural studies via a single expression screen. Protein Sci 15:1115–1126

7. Figler RA, Omote H, Nakamoto RK, Al-Shawi MK (2000) Use of chemical chaperones in the yeast *Saccharomyces cerevisiae* to enhance heterologous membrane protein expression: high-yield expression and purification of human P-glycoprotein. Arch Biochem Biophys 376:34–46

8. Sonoda Y, Cameron A, Newstead S, Omote H, Moriyama Y, Kasahara M, Iwata S, Drew D (2010) Tricks of the trade used to accelerate high-resolution structure determination of membrane proteins. FEBS Lett 584:2539–2547

9. Kota J, Gilstring CF, Ljungdahl PO (2007) Membrane chaperone Shr3 assists in folding amino acid permeases preventing precocious ERAD. J Cell Biol 176:617–628

Large-Scale Production of Membrane Proteins in *Pichia pastoris*: The Production of G Protein-Coupled Receptors as a Case Study

Shweta Singh, Adrien Gras, Cedric Fiez-Vandal, Magdalena Martinez, Renaud Wagner, and Bernadette Byrne

Abstract

One of the major advantages of using *Pichia pastoris* is that it is readily adapted to large-scale culture in bioreactors. Bioreactors allow precise regulation of cell growth parameters increasing both yields and reproducibility of the culture. *P. pastoris* cultures grow to very high cell densities which helps minimise culture volume and facilitates downstream processing of the sample. Here, we provide protocols for the large-scale production of the human adenosine A_{2A} receptor ($A_{2A}R$) and provide some details of how bioreactor cultures can be used for optimisation of expression of the human dopamine D2 receptor (D2DR).

Key words: Large-scale expression, Bioreactor, Control of culture parameters, Methanol probe, Large-scale membrane preparations, Receptor

1. Introduction

One of the key advantages of using *Pichia pastoris* is that it is readily adapted to large-scale culture in bioreactors (1, 2). Bioreactors allow precise regulation of the aeration, pH and addition of carbon source, which in turn allows the cultures to grow to high cell densities. This in turn can be used to maximise yields of the target protein. *P. pastoris* cultures are able to grow to much higher cell densities than the other popular yeast expression host, *Saccharomyces cerevisiae*. This is because at high cell densities, *S. cerevisiae* switches to fermentative growth producing ethanol as a by-product. This quickly reaches toxic levels which limits further cell growth and thus recombinant protein production. In contrast, *P. pastoris* is a relatively poor fermenter, preferring respiratory growth. It can therefore be

Roslyn M. Bill (ed.), *Recombinant Protein Production in Yeast: Methods and Protocols*, Methods in Molecular Biology, vol. 866, DOI 10.1007/978-1-61779-770-5_17, © Springer Science+Business Media, LLC 2012

grown to very high cell densities (an OD_{595} of 500 has been reported, (2)) in controlled bioreactor environments without the toxic side effects of ethanol production. This means it is possible to use much lower culture volumes for *P. pastoris* making expression cheaper as well as facilitating downstream processing.

There are a number of standard protocols available for the production of proteins in *P. pastoris* using bioreactors (3–5), many of which have been developed to deal with the specific challenges of growing high cell density cultures. However, optimisation of these standard protocols is usually necessary for specific targets. One particular issue is the osmotic stress induced during high cell density culturing, known to be responsible for adaptive cell response mechanisms, such as changes in membrane lipid content (6). These changes may be disadvantageous during membrane protein production and hence medium cell density culturing approaches may be more suitable. For example, in high cell density fermentations, we observed lowered specific activity for some recombinant G protein-coupled receptors (GPCRs) as well as altered receptor behaviour in solubilisation assays. Therefore, although it is possible to produce very high *P. pastoris* cell density cultures (up to $OD_{600} = 500$), we typically use medium cell density protocols ($OD_{600} = 80–100$; (3)).

Another concern is proteolytic degradation of the target protein, as the high stress environment of the bioreactor increases the likelihood of cell death leading to release of intracellular proteases. Our protocols therefore use a temperature-limited fed batch technique (3) which was specifically developed to reduce proteolysis during recombinant protein production. However, despite this previous work in our laboratories has revealed the proteolytic degradation of some, but not all, receptor constructs when expressed in bioreactors (7). In contrast, the same constructs are not subject to degradation in the much lower cell density cultures grown in shake flasks. This clearly illustrates the fact that specific optimisation is required for individual protein targets. For example, in one case it was necessary to express an alternative construct in order to obtain protein which was sufficiently stable for further studies (7).

A further important parameter that requires optimisation is the amount of methanol added for induction. It has been shown that very high levels of methanol can have cytotoxic effects which reduce cell viability and thus yield of the target protein (8). Methanol sensors, which detect the level of unmetabolised methanol, have been key to reducing these cytotoxic effects and optimising the production of the target protein.

The large-scale production of membrane proteins in *P. pastoris* also presents challenges following culture harvest, which are related to the structure of the yeast cell envelope (9). The cell envelope comprises (1) the plasma membrane which surrounds the cytosol and is the location for a wide range of integral membrane proteins with diverse roles in cellular function, (2) the periplasm, external to the

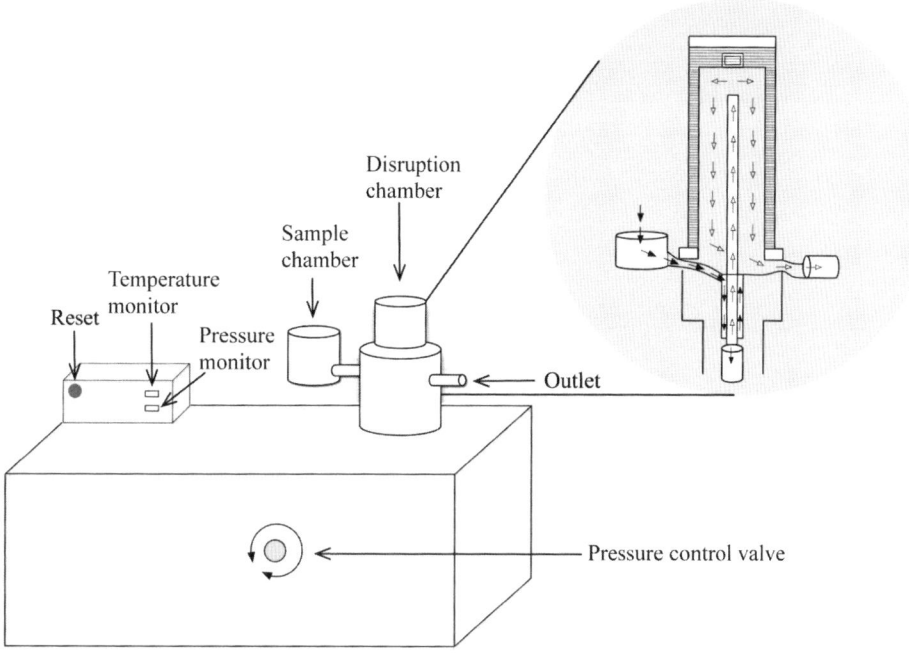

Fig. 1. A schematic diagram of the continuous cell disruptor system used for the preparation of cell membranes. Within the detailed *inset*, the *black arrows* show the flow of the cell suspension prior to cell lysis. The cells are forced through a nozzle at high pressure and then hit a target at high speed. Both processes contribute to efficient cell lysis. The cell lysate, showed by the *open arrows* is then collected from the outlet.

plasma membrane but internal to the cell wall and the location of a number of secreted enzymes with important roles in hydrolysis of substrates prior to transport across the plasma membrane and (3) the cell wall, formed from a thick layer of mainly polysaccharides with a small amount of chitin. The consequence of this complex cell envelope structure to researchers producing recombinant proteins in yeast is that specialised, and often expensive, pieces of equipment are required for efficient cell lysis of large volumes of cell pellet. Such equipment must provide large mechanical forces in order to break open the cell walls. Continuous flow cell disruptors which work by forcing cells through a small nozzle at very high pressures (Constant Systems, Fig. 1) are a popular means of processing large volumes of yeast cell culture. Bead-beaters relying on crushing and cracking of the cells with glass beads in a sealed system are also used. In both cases, it is important to perform the cell lysis at a low temperature in order to minimise proteolytic degradation of the target protein. In addition, it is possible to use enzymes to lyse cells (10); however, the effects of such enzymes on the target protein must be carefully assessed.

Below, we give a protocol for the large-scale production of the GPCR, adenosine A_{2A} receptor ($A_{2A}R$), using a bioreactor. In addition, we supply a protocol for large-scale membrane preparation using a continuous flow cell disruptor. We also provide some comments on optimising receptor yield in bioreactor cultures using the human dopamine D2 receptor (D2DR) as the target protein.

2. Materials

2.1. Medium Preparation for Large-Scale Bioreactor Cultures

1. Geneticin should be made as a stock solution at 100 mg/mL and filter-sterilised through a 0.2 μm syringe filter.

2. YPD agar: 1% (w/v) yeast extract, 2% (w/v) peptone, 2% (w/v) dextrose, 2% (w/v) agar supplemented with either 0.1 or 0.25 mg/mL geneticin. Autoclave 450 mL water containing 5 g yeast extract, 10 g peptone and 10 g agar. Add 50 mL filter-sterilised 20% dextrose to the medium once it has cooled to 60°C.

3. MGY medium: Add 100 mL autoclaved 1 M potassium phosphate buffer pH 6 to 700 mL autoclaved deionised (DI) water. Add 100 mL filter-sterilised 10× yeast nitrogen base without amino acids (to give a final concentration of 1.34% w/v), 100 mL filter-sterilised 10× glycerol solution (to give a final concentration of 1% w/v), followed by 2 mL filter-sterilised 500× biotin (to give a final concentration of 0.02% w/v).

4. FM22 medium: 0.5% (w/v) ammonium sulphate, 0.1% (w/v) calcium sulphate, 1.43% (w/v) potassium sulphate, 1.17% (w/v) magnesium sulphate, and 2% (w/v) glycerol. If a significant amount of foaming is observed, 250 μL/L of antifoam can be added to the medium before autoclaving. This solution can be autoclaved within the bioreactor vessel. Prepare 3 L of this solution.

5. Potassium phosphate solution: 1 L 4.3% (w/v) monobasic potassium phosphate, autoclaved separately to prevent precipitation. Add this to the bioreactor following autoclaving to give a final volume of 4 L.

6. PMT4 trace elements solution: 0.2% (w/v) copper sulphate, 0.008% (w/v) sodium iodide, 0.3% (w/v) manganese sulphate, 0.02% (w/v) sodium molybdate, 0.002% (w/v) boric acid, 0.05% (w/v) calcium sulphate, 0.05% (w/v) cobalt chloride, 0.7% (w/v) zinc chloride, 2.2% (w/v) iron sulphate, and 1 mL/L concentrated sulphuric acid. This solution should be stored in the dark at 4°C.

7. Biotin solution: filter sterilise a 2% solution. To increase the shelf-life of the PMT4 solution, add biotin just prior to use to a final concentration of 0.02%.

8. Glycerol feed solution: 50% (w/v) glycerol autoclaved prior to use.

9. Methanol feed solution: add 4 mL/L PMT4 solution to 400 mL 100% methanol.

10. Histidine solution: filter sterilise a 100× solution of 4% (w/v) histidine.

11. Ammonia solution: 28% ammonia solution (purchased as a 0.880 g/cm^3 solution from VWR) used to control the pH of the culture.

12. Antifoam solution: autoclave 100 mL of antifoam solution.

2.2. Large-Scale Membrane Preparation

1. Breaking buffer: 50 mM Na$_2$HPO$_4$ pH 7.5, 100 mM NaCl, 5% glycerol, 2 mM EDTA supplemented with protease-inhibitor tablets (Roche; 1 tablet/100 ml buffer).

2. Membrane buffer: 50 mM HEPES pH 7.4, 100 mM NaCl, 10% (w/v) glycerol.

3. Methods

3.1. Starter Culture

1. Select single *P. pastoris* colonies containing the plasmid of interest on separate YPD agar plates supplemented with 0.01% or 0.25% geneticin (see Note 1).

2. Use a single colony to inoculate a 150 mL MGY medium.

3. Grow the culture at 30°C with aeration for 18–22 h to an OD$_{600}$ 10–15.

3.2. Preparation of the Bioreactor

This section describes a medium density *P. pastoris* cultivation in a 5 L bioreactor vessel (4 L working volume) operated by an ADI 1010 bio-controller connected to a PC running BioExpert software (all from Applikon Biotechnology; see Fig. 2 and Note 2).

Day 1

1. Add 4 L autoclaved FM22 medium to the bioreactor vessel. Add 100 mL glycerol feed solution to a feed bottle. Connect the glycerol and antifoam (antifoam A emulsion, Sigma) feed bottles to the bioreactor using silicon tubing. Connect the empty methanol bottle using solvent-impermeable Tygon® tubing, but connect C-Flex® for the section used by the peristaltic pump (both from Masterflex). Ensure that all venting filters are protected using cotton wool and aluminium foil secured with autoclave tape. Clamp all lines except the bioreactor exhaust. Autoclave on a liquid cycle.

Day 2

2. Connect the ammonia feed bottle using Tygon® and C-Flex® tubing and add 200 mL methanol feed solution to the empty feed bottle.

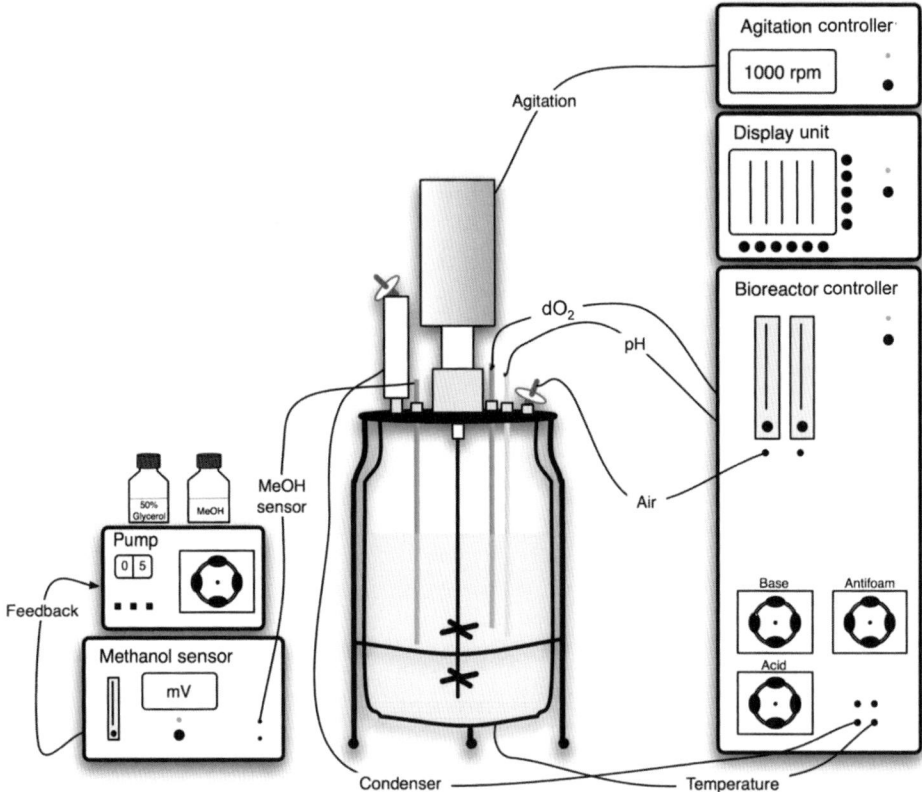

Fig. 2. A schematic diagram of the bioreactor set-up. From *left to right*: methanol sensor and pump; bioreactor vessel; and controller units. The bioreactor contains pH, dO_2, methanol, and temperature probes as well as a stirrer inserted into the vessel. These are all connected to the controller unit which allows monitoring and control of the different culture parameters. The pH is maintained at 5 by the addition of acid or base into the cell culture, as required, while the dO_2 is maintained at a minimum level of 40% by regulating the airflow. The vessel contains an external jacket connected to the condenser, which is required to maintain the temperature of the culture. During the fed-batch phase, glycerol is introduced into the system at a regular flow rate through a pump. The pump is then connected to the methanol sensor during the induction phase in order to regulate the amount of methanol added. The vessel also has inlets to add phosphate buffer, additives, and inoculum to the medium and an outlet to harvest the cells.

3. Begin equilibration of the bioreactor by setting the temperature and dO_2 control to 30°C and 35%, respectively. Add 1 mL PMT4 solution per 1 L medium at this point. Once the temperature and dO_2 have equilibrated, set the pH control to 5.0.

4. Centrifuge the MGY starter culture and resuspend the cells in 10 mL fresh MGY medium. Inoculate the bioreactor vessel with sufficient cell suspension to obtain a starting OD_{600} of 0.25.

5. Incubate the culture for 18–20 h. During this growth phase, the cells consume all the glycerol. Glycerol depletion is marked by a sharp rise in dO_2 levels (Fig. 3).

Day 3

6. Once the glycerol is fully depleted, begin the glycerol fed-batch phase by feeding 50% glycerol at 10 mL/L/h with a peristaltic

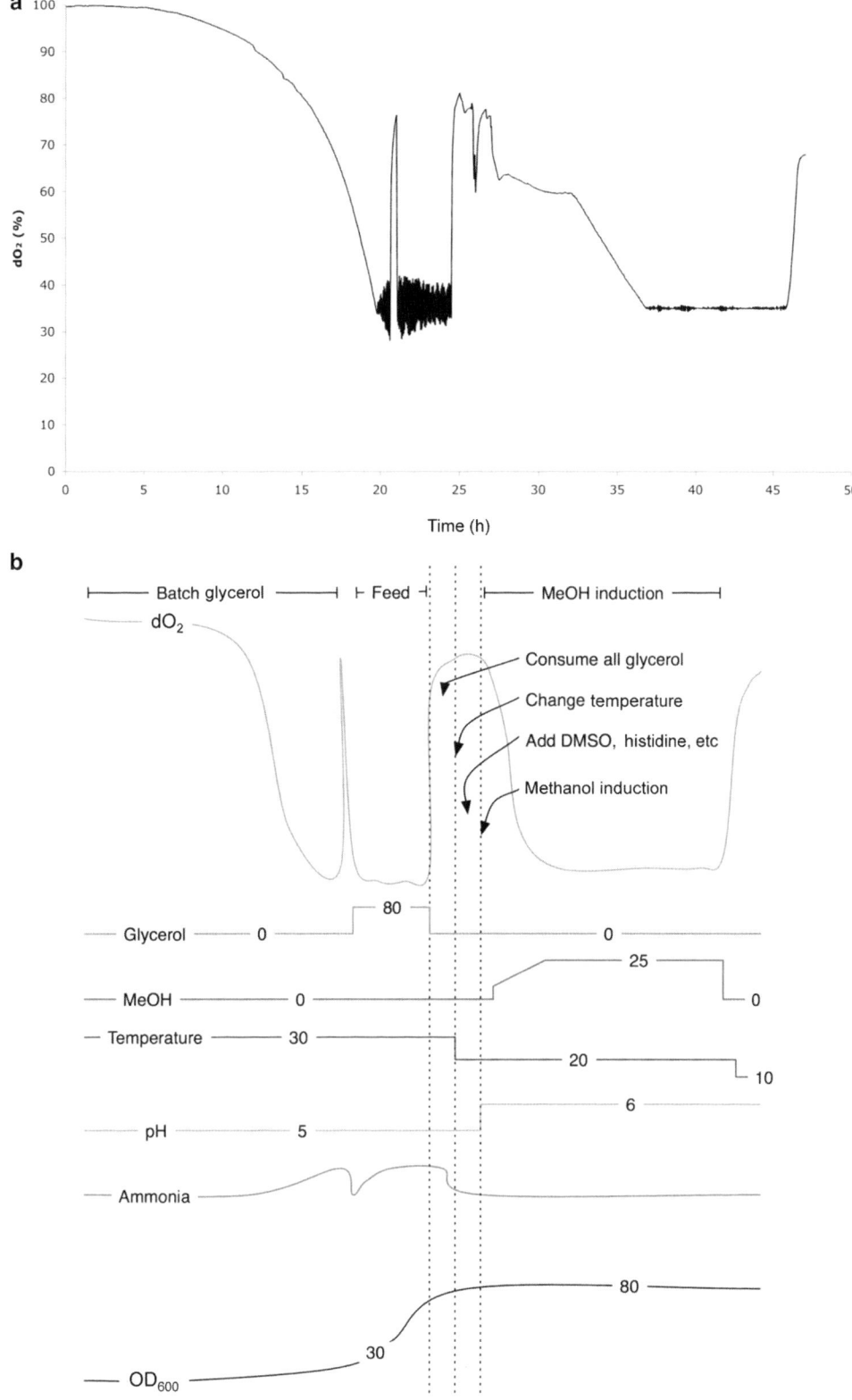

Fig. 3. (**a**) A schematic representation of the different parameters monitored and controlled during production of $A_{2A}R$ in a bioreactor. (**b**) A graphical representation of the changes in dO_2 during the course of a culture producing wild-type $A_{2A}R$. The key culture phases are indicated.

pump. Continue the feed until the OD_{600} reaches 80–100 (3–4 h).

7. Turn off the glycerol feed. A further sharp rise in dO_2 levels should be observed at this point (Fig. 3). Add sterile histidine to a final concentration of 0.04% and sterile DMSO to a final concentration of 2.5%. Allow for minor temperature changes to stabilise then lower the temperature control to 22°C.

8. Once the temperature has stabilised, begin the methanol induction phase (18–21 h) by adding 0.1% (v/v) methanol feed solution to the vessel. Allow 2–3 min for the methanol sensor (Raven Biotech) reading (mV) to stabilise and set the methanol control to this value. The internal feedback system of the methanol sensor enables the on/off control of the 101U/R peristaltic pump (Watson Marlow) to maintain a steady concentration of 0.1% methanol in the vessel. Allow approximately 1 h for the culture to adapt to methanol metabolism. The dO_2 levels should drop to 35% during this period.

9. Add another 0.1% (v/v) methanol aliquot to the vessel and set the methanol control at 0.2% as described in step 8. Repeat this step until methanol is set at 0.4%. Samples should be taken at multiple time points and frozen at −80°C for further analysis of the yield profile. In the case of recombinant $A_{2A}R$, the culture was usually induced for 18 h after which a reduction in recovery of functional receptor was observed.

Day 4

10. Turn off the methanol feed and allow dO_2 to rise to approximately 80% before harvesting cells. This allows depletion of the methanol present in the vessel. This typically takes about 30 min.

11. Spin down the cells at $3,000 \times g$ and store the pellet at −80°C or proceed directly to membrane preparation (see Notes 3 and 4).

3.3. Large-Scale Membrane Preparation

1. All steps are carried out at 4°C. Resuspend the cells in ice-cold cell breakage buffer.

2. Stir the sample gently for 15 min.

3. Pass the suspension through a continuous cell disruptor (Constant systems; Fig. 1) at 39 kpsi. To allow efficient cell breakage, the suspension is usually passed two or three times through the disruptor.

4. Remove cell debris by centrifuging samples at $5,000 \times g$ for 20 min.

5. Isolate the membranes by centrifuging the supernatant at $170,000 \times g$ for 1 h using a 45 Ti rotor in an Ultra XP ultracentrifuge (Beckman Coulter).

6. Resuspend the membranes in 400 mL membrane buffer (50 mM Tris HCl pH 8, 120 mM NaCl, 20% glycerol, 300 nM

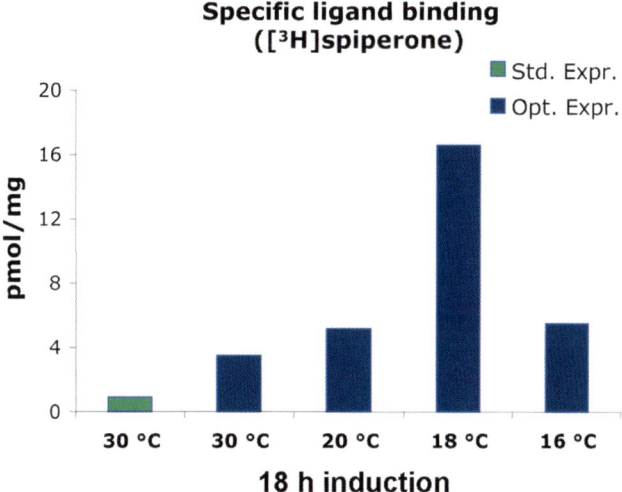

Fig. 4. Optimisation of functional expression of DRD2 in a bioreactor. Induction at 30°C for 18 h, yielded 2 pmol/mg functional membrane protein. Supplementing the culture medium with 5% DMSO, 0.04% histidine and 10 mM metoclopramide (a receptor-specific antagonist) and inducing at 30°C for 18 h gave an increased functional yield of 4 pmol/mg membrane protein. Further optimisation of the yield was achieved by reducing the induction temperature. Inducing at 20°C increased the functional yield to about 6 pmol/mg membrane protein, with a similar result obtained for induction at 16°C. The highest yield was obtained when the culture was induced at 18°C to give 16 pmol/mg membrane protein. These optimised conditions yielded an eightfold higher yield than the original induction conditions.

ZM241385 (Tocris), protease inhibitors (Roche)) using a cell homogeniser (Dounce). Use immediately for solubilisation and purification of the receptor or store at −80°C until further use (see Note 5).

3.4. Optimisation of Expression Yields in a Bioreactor

1. Prepare a starter culture and bioreactor as described above (Subheadings 3.1 and 3.2).

2. Grow the cultures as described above.

3. Induce expression at 30°C for 18 h.

4. For optimisation of D2R yields, we supplemented the culture medium with 5% DMSO, 0.04% histidine, and 10 μM metocloprimide (specific D2R antagonist).

5. Induce expression at a range of different temperatures (e.g. 20, 18, and 16°C). Figure 4 gives an example of the different yields which can be obtained by altering the culture conditions (see Note 6).

4. Notes

1. It is useful to screen the colonies using both concentrations of geneticin. Colonies in which a higher copy number of the expression plasmid has been integrated into the genome will

grow on higher concentrations of geneticin. However, there is no clear relationship between higher copy number and higher recombinant protein yields. Therefore, it is probably useful to carry out expression screening on colonies containing both high and low copy numbers.

2. The 5 L bioreactor equipment typically used in our laboratory was obtained from Applikon Biotechnology; however, comparable yields can be achieved using the BioFlo 3000 system from New Brunswick Scientific.

3. Since 350 g cells can be typically obtained from a bioreactor culture, it is usually best to freeze the cells in batches. These can then be processed separately.

4. It is useful to prepare a small (5 mL) aliquot of culture to test for production of the target protein either by Western blot or functional analysis before processing the sample further.

5. It is possible to freeze membranes containing the GPCR prior to further analysis or purification, but this can reduce functional yield. In most cases, it is preferable to prepare membranes immediately prior to use.

6. Although optimisation of yields in a bioreactor is possible, it may also be quicker to assess the effects of different conditions in shake flasks.

Acknowledgments

This research was funded by the MepNet consortium, the Biotechnology and Biological Research Council and GlaxoSmithKline.

References

1. Cereghino GPL, Cereghino JL, Ilgen C, Cregg JM (2002) Production of recombinant proteins in fermenter cultures of the yeast *Pichia pastoris*. Curr Opin Biotechnol 13:329–332

2. Cereghino JL, Cregg JM (2000) Heterologous protein expression in the methylotrophic yeast *Pichia pastoris*. FEMS Microbiol Rev 24:45–66

3. Jahic M, Wallberg F, Bollok M, Garcia P, Enfors S-O (2003) Temperature limited fed-batch technique for control of proteolysis in *Pichia pastoris* biorector cultures. Microb Cell Fact 2:6–17

4. Charoenrat T, Ketudat-Cairns M, Stendahl-Andersen H, Jahic M, Enfors S-O (2005) Oxygen-limited fed-batch process: an alternative control for *Pichia pastoris* recombinant protein processes. Bioprocess Biosyst Eng 27:399–406

5. Jahic M, Veide A, Charoenrat T, Teeri T, Enfors S-O (2006) Process technology for production and recovery of heterologous proteins with *Pichia pastoris*. Biotechnol Prog 22:1465–1473

6. Mattanovich D, Gasser B, Hohenblum H, Sauer M (2004) Stress in recombinant protein producing yeasts. J Biotechnol 113:121–135

7. Singh S, Gras A, Fiez-Vandal C, Ruprecht J, Rana R, Martinez M, Strange PG, Wagner R, Byrne B (2008) Large-scale functional expression of WT and truncated human adenosine A2A receptor in *Pichia pastoris* bioreactor cultures. Microb Cell Fact 7:28–38

8. Zhang W, Smith LA, Plantz BA, Schlegel VL, Meagher MM (2002) Design of methanol feed control in *Pichia pastoris* fermentations based upon a growth model. Biotechnol Prog 18:1392–1399

9. Northcote DH, Warne RW (1952) The chemical composition and structure of the yeast cell wall. Biophys J 51:232–236.2

10. Salazar O, Asenjo JA (2007) Enzymatic lysis of microbial cells. Biotechnol Lett 29:985–994

Large-Scale Production of Membrane Proteins in *Saccharomyces cerevisiae*: Using a Green Fluorescent Protein Fusion Strategy in the Production of Membrane Proteins

David Drew and Hyun Kim

Abstract

The production of membrane proteins in the large quantities necessary for structural analysis requires many optimization steps. The GFP-fusion-based scheme described in earlier chapters (Chapters 4, 8, and 16) facilitates these steps by allowing the selection of high yielding clones that produce detergent-stable membrane proteins. Here, we describe the experimental steps required to establish the reproducible, large-scale production and purification of membrane protein–GFP fusions using *S. cerevisiae*.

Key words: Membrane protein, Green fluorescent protein, Overexpression, *Saccharomyces cerevisiae*, Fluorescence-detection size exclusion chromatography

1. Introduction

To facilitate the high-throughput screening and purification of membrane proteins for structural and functional studies, we have developed a protocol based on fluorescent monitoring of a carboxy-terminal GFP fusion tag (1–6). Analysis of its fluorescence enables the yields of membrane protein–GFP fusions to be easily estimated in both whole cells and isolated membranes. Further, this approach facilitates the optimization of membrane protein–GFP fusion yields, as well as detergent screening and purification by fluorescence-detection size exclusion chromatography (FSEC) (7). Using this approach, membrane protein–GFP fusions that are well expressed and stable in detergent, as judged by their monodispersity by FSEC, can be selected for downstream large-scale production and purification (Fig. 1).

Roslyn M. Bill (ed.), *Recombinant Protein Production in Yeast: Methods and Protocols*, Methods in Molecular Biology, vol. 866, DOI 10.1007/978-1-61779-770-5_18, © Springer Science+Business Media, LLC 2012

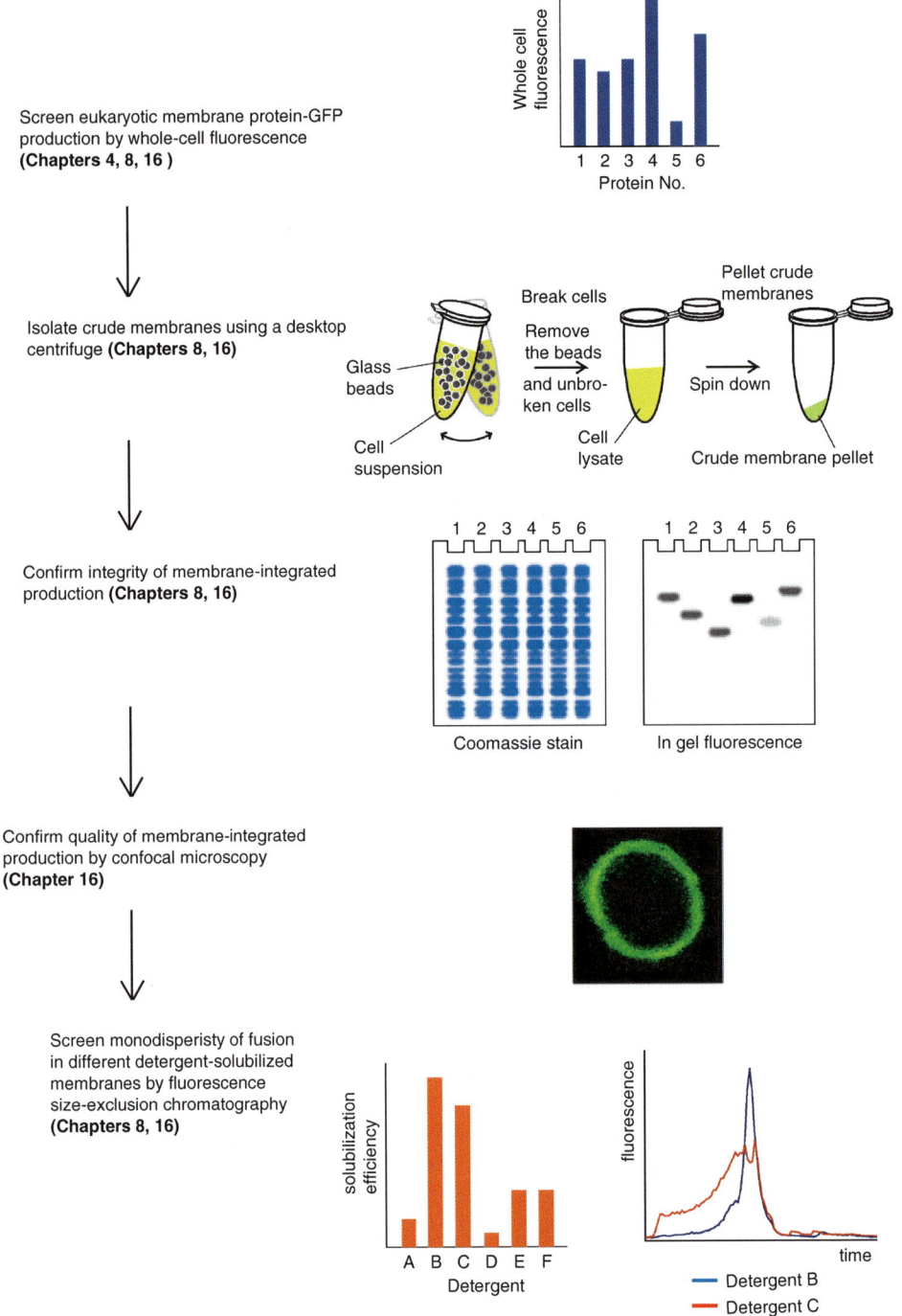

Fig. 1. Flowchart illustrating the screening process for the production and purification of eukaryotic membrane protein–GFP fusions in *Saccharomyces cerevisiae* (Figure adapted from ref. 5). Most targets are evaluated based on the quality of the FSEC trace in DDM.

This chapter describes a protocol for (1) the large-scale isolation of membranes, (2) purification of membrane protein–GFP fusions and (3) cleavage of the GFP-His$_8$ tag from membrane protein–GFP fusions.

2. Materials

1. Medium lacking uracil: 2 g yeast synthetic drop-out medium without uracil, 6.7 g yeast nitrogen base without amino acids and either 2% glucose (for pre-culture) or 0.1% glucose (expression culture), adjusted to a final volume of 1 L. For plates, use 2% glucose and add 20 g bacto agar.

2. 96-well black optical-bottom plates (Nunc).

3. SpectraMax M2e microplate reader (Molecular Devices).

4. Benchtop ultracentrifuge, Beckman Coulter Optima MAX series with TLA-55 and TLA 120.1 rotors (Beckman).

5. 1.5 mL polyallomer microcentrifuge tubes (Beckman).

6. Imidazole, minimum 99% (Sigma).

7. Ni-NTA superflow resin (Qiagen).

8. His-tagged TEV protease (8), stored at –80°C in buffer containing 50% glycerol, 20 mM Tris–HCl, pH 7.5, 0.3 M NaCl, and 1 mM DTT.

9. Cell resuspension buffer (CRB): 50 mM Tris–HCl, pH 7.6, 1 mM EDTA, 0.6 M sorbitol.

10. Constant Systems TS series cell disruptor (Constant Systems).

11. Membrane resuspension buffer (MRB): 20 mM Tris–HCl, pH 7.6, 0.3 M sucrose, 0.1 mM CaCl$_2$.

12. Bicinchoninic acid (BCA) protein assay kit (Pierce).

13. Phosphate-buffered saline (PBS): 1.44 g Na$_2$HPO$_4$·2H$_2$O (8.1 mM phosphate), 0.25 g KH$_2$HPO$_4$ (1.9 mM phosphate), 8 g NaCl, 0.2 g KCl, adjust pH to 7.4 using 1 M NaOH or HCl and adjust final volume to 1 L.

14. Equilibration buffer (EB): 1 × PBS, 10 mM imidazole pH 8.0, 150 mM NaCl, 10% glycerol (w/v), 3 × critical micelle concentration (CMC) detergent of choice.

15. Dialysis buffer (DB): 20 mM Tris–HCl, pH 7.5, 0.15 M NaCl, 3 × CMC detergent of choice.

16. 5 mL His-trap columns (GE Healthcare).

17. Dialysis tubing, 12–14 kDa molecular weight cut-off (Spectrum labs).

18. Poly-Prep glass Econo-Column chromatography columns (Bio-Rad).

19. Superdex 200 10/300 GL Tricorn gel filtration column (GE Healthcare).

20. Centrifugal filter devices (Millipore/Vivascience).

21. 15 L or 50 L bioreactor vessels (Applikon).

22. ÄKTA FPLC system (GE Healthcare).

23. Frac-950 fraction collector with rack C (GE Healthcare).

24. Peristaltic pump P-1 (GE Healthcare).

3. Methods

3.1. Large-Scale Isolation of Membranes

1. Inoculate 10 mL medium lacking uracil containing 2% glucose with transformed yeast cells and incubate overnight in an orbital shaker at 280 rpm and 30°C. The following day, transfer the overnight culture to a 500-mL shake flask containing 150 mL medium lacking uracil with 2% glucose and incubate overnight as in the first step.

2. Dilute the 150 mL overnight culture to an OD_{600} of 0.12 into 1 L medium lacking uracil containing 0.1% glucose in a 2.5 L baffled shake flask. Incubate the culture in an orbital shaker at 280 rpm and 30°C. Induce at OD_{600} 0.6 using the parameters established in the initial optimization screens (Chapters 8 and 16). Use 10–15 L flasks for large-scale production. Alternatively, 15 L or 50 L bioreactors can be used to obtain similar yields per cell to shake flasks.

3. After a 22 h incubation, harvest the cells by centrifugation at $4,000 \times g$ at 4°C for 10 min. Decant the supernatant and resuspend the cell pellet in 25 mL of cell resuspension buffer (CRB) per L original cell culture.

4. Break cells using a heavy-duty cell disrupter for four passes at incremental pressures of 25, 30, 32, and 35 kpsi (1.7–2.4×10^3 atm) at 4–15°C. Remove 100 μL cells, transfer to a 96-well plate and measure GFP fluorescence emission at 512 nm and excitation at 488 nm in a microplate spectrofluorometer.

5. Remove unbroken cells and debris by centrifugation at $10,000 \times g$ at 4°C for 10 min and collect the supernatant, which contains the membranes. Transfer 100 μL supernatant to a 96-well plate and measure GFP fluorescence. Calculate the yeast cell breakage efficiency by comparing GFP fluorescence to that measured in step 4 (see Note 1).

6. To collect the membranes, centrifuge the cleared supernatant at $150,000 \times g$ at 4°C for 120 min. Discard the supernatant

and resuspend the pellet to a final volume of 6 mL/L of original cell culture with MRB using a disposable 10 mL syringe with a 21 gauge needle. Transfer 100 μL membrane suspension to a 96-well plate and measure GFP fluorescence as outlined in step 4. Calculate the yield of membrane protein based on this fluorescence reading as well as the total protein concentration using the BCA protein assay kit following the manufacturer's instructions.

7. Freeze membranes in liquid nitrogen and store at –80°C (see Note 2).

3.2. Purification of Membrane Protein–GFP Fusions (Fig. 2)

1. Dilute the membrane suspension isolated from 10 to 15 L of culture into a 500 mL beaker to a final protein concentration of 3 mg/mL using equilibration buffer (EB). Add the detergent powder that produced the best FSEC trace (see Chapter 8) to a final concentration of 1 or 2% (w/v) depending on the CMC of the selected detergent. Incubate the mixture at 4–10°C for 1 h and use a magnetic stirrer to mix the solution.

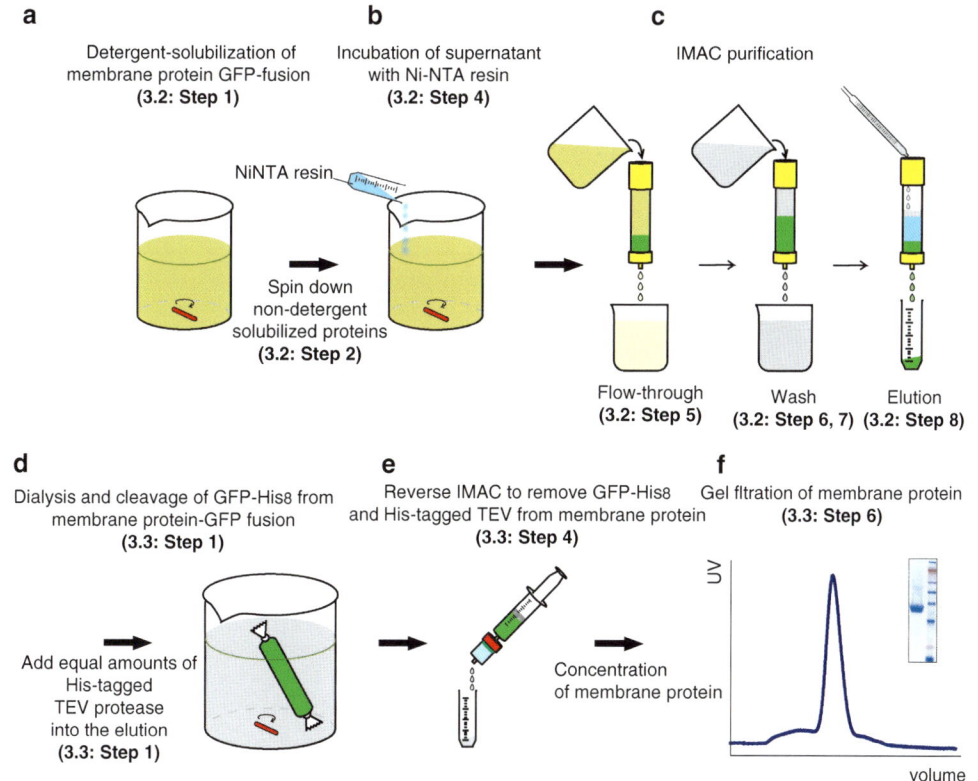

Fig. 2. Flowchart illustrating the purification of eukaryotic membrane proteins from GFP fusions (figure adapted from ref. 5). This protocol follows a standard Ni-NTA purification procedure. However, we recommend dialysis for efficient TEV cleavage and passing material through a higher capacity Ni-NTA column than in the first step (using the His-Trap™ column) to obtain very pure protein.

2. Pellet the insoluble material by centrifugation at $100,000 \times g$ at 4°C for 45 min. Transfer the supernatant into an appropriately sized glass beaker. Transfer 100 µL detergent-solubilized membranes to a 96-well plate and measure GFP fluorescence as outlined in Subheading 3.1, step 4. Calculate the amount of GFP fusion as detailed in Chapter 8.

3. Use 1 mL Ni-NTA resin (2 mL of a 50% slurry) per milligram of GFP fusion to be purified (see Subheading 3.2, step 2) and equilibrate with five column volumes of EB (see Note 3).

4. Add the equilibrated Ni-NTA resin to detergent-solubilized membranes. Use a magnetic stirrer to mix the solution at 4°C for 2–3 h (see Note 4).

5. Pour the slurry into a glass Econo-Column. Transfer 100 µL of the flow-through to a 96-well plate and measure GFP fluorescence as outlined in Subheading 3.1, step 4. Compare with the reading taken in Subheading 3.2, step 2 and calculate the binding efficiency of the fusion protein to the Ni-NTA resin.

6. Wash the column with 10 column volumes of EB.

7. Add 1 M imidazole to EB to a final concentration of 30 mM and wash the column with 35 column volumes.

8. Add 1 M imidazole to EB to a final concentration of 250 mM and elute the protein from the column using 2 column volumes.

9. Transfer 100 µL eluate to a 96-well plate and measure GFP fluorescence as outlined in Subheading 3.1, step 4. Determine the amount of GFP fusion as outlined in Chapter 8. The amount of GFP fusion in the eluate should be determined by measuring the GFP fluorescence. The BCA assay measures total protein, including any contaminants.

3.3. Removal of the GFP-His$_8$ Tag from Membrane Protein–GFP Fusions

1. The total amount of membrane protein–GFP fusion is calculated as described in Chapter 8. For every 1 mg of membrane protein–GFP fusion, add 1 mg of tobacco etch virus (TEV) protease. The total reaction volume is typically 40 mL. Transfer the reaction mix to dialysis tubing and dialyze overnight at 4°C against 3 L of dialysis buffer (DB).

2. After overnight dialysis, inject 500 µL digested material onto a size exclusion chromatography column (SEC) and confirm by FSEC that all protein has been cleaved by the His-tagged TEV protease (see Note 5).

3. To remove the His-tagged TEV protease, cleaved GFP-His$_8$ tag and coeluting contaminating proteins from the reaction mix, equilibrate a 5 mL His-trap column with DB.

4. Pass the dialyzed sample through the His-Trap column at a flow rate of 2 mL/min. Collect the flow-through, which should contain the target membrane protein, into a 50 mL falcon

tube. If the membrane protein lacking a His$_8$-tag still binds to the resin, wash with 30 mM imidizole and collect the flow-through. Elute the bound material, which contains the cleaved GFP and His-tagged TEV protease, and analyze all fractions by SDS-PAGE (see Note 6).

5. Concentrate the flow-through from step 4 to 0.5 mL using a centrifugal concentrator. Calculate the total yield of protein using the BCA assay and adjust the concentration to 20 mg/mL, if necessary. Inject 0.5 mL onto a Superdex 200 10/300 column pre-equilibrated with the buffer used for dialysis at a flow rate of 0.5 mL/min.

6. Collect 0.5 mL fractions. Analyze the UV trace and pool the fractions containing the membrane protein. For structural work, concentrate using a 100 molecular weight cut-off (MWCO) concentrator to avoid concentrating detergent micelles, as these can inhibit crystallization (see Note 7).

4. Notes

1. Breakage efficiency should be around 80%. If lower, dilute the cells further with buffer as this increases the efficiency of breakage. Depending on the stability of the recombinant membrane protein, incubate cells with zymolase, which removes the cell wall, prior to breakage.

2. If the membranes isolated from 1 L of cells are resuspended in 6 mL MRB, the GFP fluorescent counts typically match the original whole-cell fluorescent counts. This corresponds to approximately 60% of the amount of GFP measured in whole cells, which is the fraction recovered in membranes.

3. It is worth mentioning that as the His$_8$-tag is carboxy-terminal to GFP, the amount of GFP that binds as a fusion is calculated at this step rather than the amount of membrane protein. Although the amount of resin used is much higher than for soluble proteins we have empirically found that this amount is necessary to ensure a good binding efficiency.

4. Batch binding of yeast-solubilized membranes improves protein recovery compared to flow loading. A 1 mL sample can be pelleted to compare fluorescence with that measured in step 2 of Subheading 3.2. Usually, binding efficiency reaches a plateau after 2 h, with only a modest gain of 10–15% after overnight incubation.

5. The most common reason for incomplete cleavage is that insufficient TEV protease is used. We typically cleave in the presence of 1 mM DTT. However, EDTA is not added as this may interfere with subsequent Ni-NTA chromatography steps.

6. When the purification is carried out for the first time, it is important to verify that the protein does not bind to the column. Once this is established, it is possible to proceed directly to the next step in all future purifications.

7. Note the concentrator flow-through should be retained in case the membrane protein–detergent complex passes through.

Acknowledgments

This work was supported by the Royal Society (United Kingdom) through a University Research Fellowship to DD and by a Basic Science Research Program grant through the National Research Foundation of Korea (NRF) funded by the Ministry of Education, Science and Technology (NRF0409-20100093) to HK.

References

1. Sonoda Y, Cameron A, Newstead S, Omote H, Moriyama Y, Kasahara M, Iwata S, Drew D (2010) Tricks of the trade used to accelerate high-resolution structure determination of membrane proteins. FEBS Lett 584:2539–2547

2. Drew DE, von Heijne G, Nordlund P, de Gier JW (2001) Green fluorescent protein as an indicator to monitor membrane protein overexpression in *Escherichia coli*. FEBS Lett 507:220–224

3. Drew D, Slotboom DJ, Friso G, Reda T, Genevaux P, Rapp M, Meindl-Beinker NM, Lambert W, Lerch M, Daley DO, Van Wijk KJ, Hirst J, Kunji E, De Gier JW (2005) A scalable, GFP-based pipeline for membrane protein overexpression screening and purification. Protein Sci 14:2011–2017

4. Drew D, Lerch M, Kunji E, Slotboom DJ, de Gier JW (2006) Optimization of membrane protein overexpression and purification using GFP fusions. Nat Methods 3:303–313

5. Drew D, Newstead S, Sonoda Y, Kim H, von Heijne G, Iwata S (2008) GFP-based optimization scheme for the overexpression and purification of eukaryotic membrane proteins in *Saccharomyces cerevisiae*. Nat Protoc 3: 784–798

6. Newstead S, Kim H, von Heijne G, Iwata S, Drew D (2007) High-throughput fluorescent-based optimization of eukaryotic membrane protein overexpression and purification in *Saccharomyces cerevisiae*. Proc Natl Acad Sci USA 104:13936–13941

7. Kawate T, Gouaux E (2006) Fluorescence-detection size-exclusion chromatography for precrystallization screening of integral membrane proteins. Structure 14:673–681

8. Lucast LJ, Batey RT, Doudna JA (2001) Large-scale purification of a stable form of recombinant tobacco etch virus protease. Biotechniques 30:544–546

Chapter 19

Large-Scale Production of Secreted Proteins in *Pichia pastoris*

Nagamani Bora

Abstract

The production of recombinant therapeutic proteins is an active area of research in drug development. These bio-therapeutic drugs target nearly 150 disease states and promise to bring better treatments to patients. However, if new bio-therapeutics are to be made more accessible and affordable, improvements in production performance and optimization of processes are necessary. A major challenge lies in controlling the effect of process conditions on production of intact functional proteins. To achieve this, improved tools are needed for bio-processing. For example, implementation of process modeling and high-throughput technologies can be used to achieve quality by design, leading to improvements in productivity. Commercially, the most sought after targets are secreted proteins due to the ease of handling in downstream procedures. This chapter outlines different approaches for production and optimization of secreted proteins in the host *Pichia pastoris*.

Key words: Secreted protein, *Pichia pastoris*, Induction, Promoter, Vector, Bioreactor

1. Introduction

Secreted proteins account for nearly one-tenth of the human genome. They function in signaling pathways, blood coagulation, immune defense, digestive processes, and comprise components of the extracellular matrix. They are also potential drug targets with recombinant examples including antibodies, erythropoietin, insulin, interferon, plasminogen activators, growth hormone, and colony-stimulating factors. Many of these proteins can be produced recombinantly in yeast (1) or bacteria (2). Classically, secreted proteins are characterized by an amino-terminal signal sequence which mediates protein translocation into the endoplasmic reticulum (ER) and secretory pathway. N-linked glycosylation on the Asn-X-Ser/Thr motif and disulfide bond formation are subsequently accomplished in the ER before exiting into the cytoplasm.

Roslyn M. Bill (ed.), *Recombinant Protein Production in Yeast: Methods and Protocols*, Methods in Molecular Biology, vol. 866, DOI 10.1007/978-1-61779-770-5_19, © Springer Science+Business Media, LLC 2012

1.1. Pichia pastoris as a Host for the Production of Secreted Proteins

The methylotrophic yeast *Pichia pastoris* was initially used for the production of single-cell protein. This was because of its ability to grow to very high cell densities in simple defined medium containing methanol, which was a cheap carbon source prior to the 1970s. Since then significant advances in the commercial development of new strains and vectors has led to a better understanding of the biology of the *Pichia* species. *P. pastoris* has therefore emerged as a specialist host system for producing recombinant proteins and has achieved widespread use due to its launch in kit form by Invitrogen (San Diego, USA).

As a eukaryotic organism, it has many of the advantages of higher eukaryotic expression systems including its ability to perform posttranslational modifications such as protein processing, proper folding, disulphide bridge formation, and protein secretion into the medium. This latter attribute later facilitates purification (1, 3). It is, however, much less expensive as a system than mammalian cell-lines and can be scaled up much more readily than *E. coli* on account of its ability to generate high biomass yields. Several examples, including the production of 12 g/L tetanus toxin fragment C and 3 g/L secreted human serum albumin, show that *P. pastoris* can routinely achieve high yields (accounting for 5–40% cell protein) when compared with bacterial and mammalian expression systems. All these features make *P. pastoris* a very useful host system. With the availability of different host strains, vectors, and chromosomal sites of integration, there is great scope for optimizing yields. This chapter attempts to give an overview of the protocols available for optimization of the production of secreted proteins in *P. pastoris*, as summarized in Fig. 1.

1.2. Small-Scale Screening for High-Yielding Clones

The identification of a high-yielding clone is central to establishing a successful protein production protocol. Clonal variations mean that at least 20 clones must typically be screened for subsequent scale-up. Since *P. pastoris* can grow under a wide range of pH, dissolved oxygen (DO), and temperature conditions, tracking the optimal combination that results in high recombinant protein yields is a primary target to reduce process development time. The medium composition also influences productivity (see Chapter 1) and hence screening in both minimal and complex media provides valuable information about the process in scale-up. Recently, protein production under low DO conditions has proved to be an effective strategy (4). A "design of experiments" (DoE) approach that involves multiple experiments done simultaneously is the best way forward to assess the impact of various input parameters (see Chapter 11).

Unfortunately, the development of high-throughput tools for the manipulation of biological systems has lagged behind conventional, low-throughput instrumentation which tends to be expensive, labor intensive, and complex. For example, bench- and large-scale

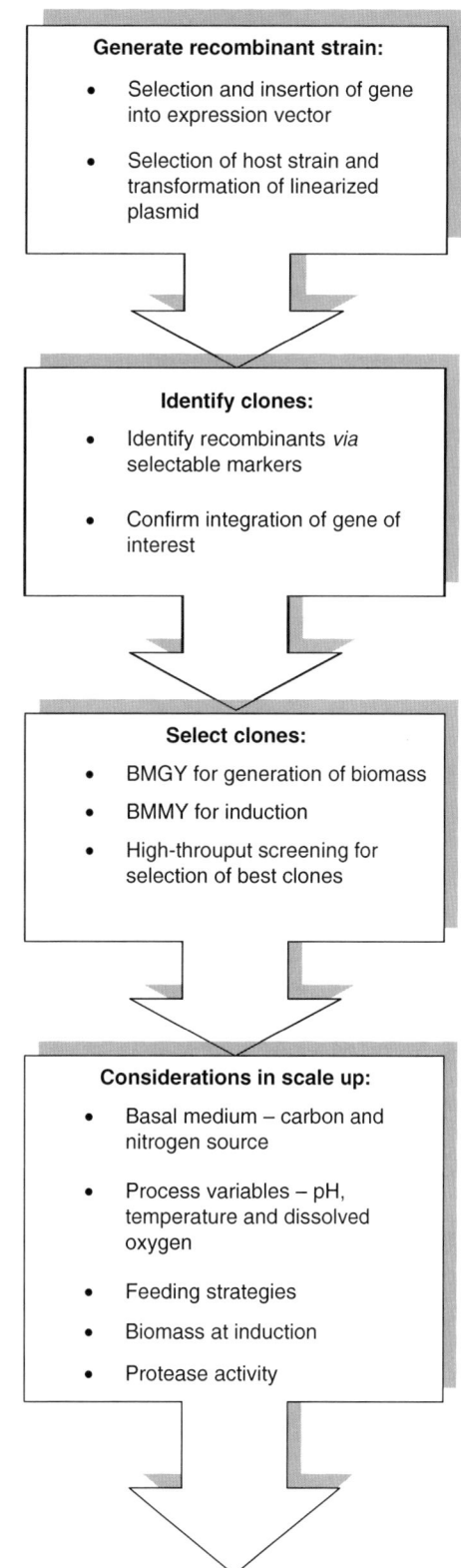

Fig. 1. Flow chart giving an overview of a typical protein production strategy for a secreted recombinant protein in *Pichia pastoris.*

Table 1
Expression vectors used for secreted proteins

Name	Selectable marker	Feature
pHIL-SI	*HIS4 amp^R*	*AOX1* promoter *PHO1* secretion signal
pPIC9K	*HIS4 kan^R*	*AOX1* promoter α-MF secretion signal
pPICZα	*Zeo^R*	*AOX1* promoter α-MF secretion signal
pGAPZα	*Zeo^R*	*GAP* promoter α-MF secretion signal

bioreactors, which control temperature, pH, and DO levels, are routinely used to produce valuable data on a protein production experiment of interest. However, bioreactors are not suitable for high-throughput optimization studies of growth and yield as they are time- and labor-intensive. Small-scale solutions are therefore emerging to address these challenges. Pall Life Sciences has launched one such system which is composed of 24 miniature parallel bioreactors. Each bioreactor has a 10 mL volume, with working volumes of 3–7 mL. This system provides automated control of pH, temperature, and DO. Substrate addition is done manually in an aseptic manner. The set-up provides a robust way of screening for suitable clones compared with standard multi-well plate or shake-flask-based approaches.

1.3. Optimization Strategies for Scaling-up the Production of Secreted Proteins

Improving the productivity of a recombinant protein production experiment is possible using different promoter and host strain combinations (5). Examples of host strains and expression vectors are given in Tables 1 and 2. Experiments in wild-type strains compared with auxotrophic (e.g., GS115; Table 2) or protease-deficient strains (e.g., SMD1163 and SMD1168; Table 2) can give a clear indication of the influence of host metabolism on recombinant protein yields.

1.3.1. Host Strains

1.3.2. Inducible Promoters

The expression of a gene of interest in *P. pastoris* is most commonly achieved using the methanol-inducible alcohol oxidase promoter system (*AOX*), which is one of the strongest and most tightly regulated promoters in this yeast species. Glycerol is the preferred carbon source as it permits derepression and not expression of *AOX* genes, while glucose represses their expression even in the presence of methanol (see Chapter 15; (6)). Based upon their methanol utilization phenotype, three types of host strain can be identified (Tables 2 and 3): a wild-type or methanol-utilization-plus phenotype

Table 2
***Pichia pastoris* strains**

Strain	Genotype	Phenotype
X-33	wild-type	Mut⁺His⁺
GS115	*his4*	Mut⁺ His⁻
KM71	*arg4 his4 aox1::SARG4*	Mutˢ His⁻ Arg⁺
SMD1163	*his4 pep4 prb1*	Mut⁺ His⁻, proteinases A, B and carboxypeptidase Y deficient
SMD1165	*his4 prb1*	Mut⁺ His⁻, proteinase B deficient
SMD1168	*his4 pep4*	Mut⁺ His⁻, proteinase A and carboxypeptidase Y deficient; partial reduction in proteinase B activity
MC100-3	*aox1Δ::SARG4 aox2Δ::Phis4 his4 arg4*	Mut⁻ His⁻

Table 3
Phenotype of clones in different host strains

Restriction enzyme	Integration event	GS115 phenotype	KM71 phenotype
Sal I or *Stu* I	Insertion at *HIS4*	His⁺ Mut⁺	His⁺ Mutˢ
Sac I	Insertion at 5′ *AOX1* region	His⁺ Mut⁺	His⁺ Mutˢ

(Mut⁺); a slow-methanol-utilization phenotype, which is due to the disruption of the *AOX1* gene (and which accounts for 90% of the alcohol oxidase enzyme activity in the cell; Mutˢ); and a methanol-utilization-negative phenotype, which is due to the disruption of both *AOX1* and *AOX2* genes (the latter accounts for 10% of the alcohol oxidase enzyme activity in the cell). The Mutˢ phenotype can sometimes be beneficial as it is not sensitive to residual methanol in the culture medium, although it is widely noted for its low specific productivities.

Other systems exist in addition to the methanol-inducible ones described above. For example, the glyceraldehyde-3-phosphate dehydrogenase (*GAP*) promoter system is used for constitutive expression of genes encoding secreted proteins. This system is strongly regulated by glucose or glycerol, but not methanol (7). It is most suited for large-scale production, as it is cost-effective and less hazardous than using methanol-induced processes.

Characterization of the formaldehyde dehydrogenase gene, which plays an important role in the catabolism of methanol and the metabolism of methylamines, led to development of the formaldehyde dehydrogenase (*FLD*) promoter system (8). *FLD* promoters are induced by methanol as a carbon source or methylamine as a nitrogen source. This system can also be used efficiently with carbon sources such as sorbitol or nitrogen sources such as ammonia. It has been shown to be useful in high cell density fermentations mimicking methanol-based processes (8).

1.3.3. Culture Parameters

P. pastoris has the advantage of being able to grow under a wide range of pH (3–7), temperature (15–32°C), and DO (5–40%) conditions. Process optimization within these ranges differs from protein to protein, depending on the nature of the protein being secreted, and is strictly process specific. Traditional methods of changing one variable at a time to achieve the best productivity are expensive and time consuming. The concept of design-of-experiments (DoE) allows different inputs for each process variable to be examined in combination, while giving valuable information on the interactions between these different input parameters (Chapter 11; (9)). Experimental designs are focused on generating a predictive equation that can be used to identify the highest-yielding process conditions (10).

Controlling specific growth rates post-induction is crucial to sustaining productivity in the secretion of heterologous proteins. Mathematical modeling of such processes, to maintain constant secretion rates, has been in use since the 1970s in both continuous and fed-batch mode (11). Furthermore, a more structured model can be constructed if the physiological state of the cell is constant, which is possible in chemostat cultivations. Modeling *P. pastoris* growth with a different carbon source pre- and post-induction is widely used, e.g., the regulation of the methanol feed post-induction to ensure the concentration is below 5 g/L in the medium. This careful balancing act benefits from modeling, which allows trial and error simulations (12).

Addition of substrates post-induction will influence the host metabolism and subsequent recombinant protein yields. Several feeding strategies, including the addition of secondary carbon sources such as glycerol (6) and sorbitol (13), have been used to increase the productivity of a process. Sorbitol has the advantage of being a non-repressive carbon source when used with methanol. Simultaneous accumulation of biomass and product can therefore be achieved by using different sorbitol: methanol ratios. A ratio in the region of 60:40 helps in controlling heat production from methanol by maintaining growth on sorbitol. Methanol concentrations can be maintained below toxic levels (5 g/L) using probes such as the Raven Biotech Inc. methanol sensor system, as part of a feedback loop.

Using glycerol with methanol as a mixed feed is also known to increase productivity, even though glycerol represses the *AOX* promoter. This is probably because the transition phase between the two carbon sources can be efficiently controlled by the cell (14). To enable this, careful monitoring of residual glycerol and methanol is used to increase titers of the target protein. Kinetic modeling of cell growth under carbon-limited conditions can ensure consumption of all of the glycerol and methanol either in continuous or fed-batch mode (15). Glycerol: methanol ratios of 4:1, 2:1, and 1:1 have all been used successfully to maintain residual methanol between 0.15 and 0.55% (15) although substrate utilization under production conditions must be monitored on a case-by-case basis.

1.3.4. Minimizing Protease Activity

Proteins secreted by *P. pastoris* are prone to proteolysis as a result of stress responses, irregular growth rates, medium imbalances, and inadequate feed profiles (16). Vacuolar proteases have the most significant impact, especially in high cell density cultivations, together with cell lysis (16). Several strategies are available to circumvent this problem, for instance using readily available protease deficient host strains, changing the process parameters at the point of induction and adding supplements to the medium. The host strains in Table 2, which are deficient in proteases A, B and carboxypeptidase Y, are known to be effective in reducing proteolysis (16).

2. Materials

2.1. Plasmid Linearization and Purification

1. Plasmid containing the gene of interest (a choice of host vectors is given in Table 1).
2. Restriction enzyme, *Pme*I, *Sac*I, or *Bst*XI, depending on the sequence of the gene; use the enzyme which does not cut the sequence of interest.
3. Water bath at 37°C.
4. Agarose, ethidium bromide, and gel electrophoresis running buffer (1× Tris acetate or Tris borate) for checking linearization of the plasmid.
5. Phenol: chloroform: isoamylalcohol (25:24:1 v/v).
6. 100% ethanol (ice cold).
7. 70% ethanol (ice cold).
8. 3 M sodium acetate.
9. Bench-top centrifuge.

2.2. Preparation of Competent Cells

1. YPD (yeast extract peptone dextrose medium): 1% yeast extract, 2% peptone, 2% dextrose (glucose). Prepare 10× glucose

(20% w/v), filter sterilize and add aseptically to the autoclaved medium. For YPD agar, add 2% agar.

2. Single colony from a YPD agar plate. A list of host strains given in Table 2.

3. BEDS solution with DTT: 10 mM bicine pH 8.3, 3% (v/v) ethylene glycol, 5% (v/v) DMSO (dimethyl sulfoxide), 1 M sorbitol supplemented with 1 mL 1.0 M DTT (dithiothreitol).

4. BEDS solution without DTT: 10 mM bicine pH 8.3, 3% (v/v) ethylene glycol, 5% (v/v) DMSO, 1 M sorbitol.

5. Bench-top centrifuge.

6. Shaking incubator at 30°C.

7. Spectrophotometer.

2.3. Transformation of the Plasmid into Host Cells

1. 1 M sterile sorbitol (stored at 4°C).

2. YPD medium (stored at 4°C).

3. YPDS agar plates supplemented with the appropriate antibiotic (e.g. 100–2,000 µg/mL zeocin): 1% yeast extract, 2% peptone, 2% dextrose, 1 M sorbitol, 2% bacteriological agar. Add the glucose, as in Subheading 2.2. Add the antibiotic once the medium has cooled to about 50°C.

4. Competent cells (stored at –80°C).

5. Electroporator (1,800 V, 15 ms pulse length).

6. Electroporation cuvettes (stored at 4°C).

7. Ice.

2.4. Screening for Integrants

1. Thermal cycler.

2. Gel electrophoresis equipment.

3. PCR kit (Stratagene) including 10× Taq DNA polymerase reaction buffer (500 mM KCl; 100 mM Tris–HCl pH 9.0; 25 mM $MgCl_2$; 1% v/v Triton X-100).

4. Forward and reverse primers targeting the host promoter: *AOX1* 5′ primer (5′-GACTGGTTCCAATTGACAAGC-3′); *AOX1* 3′ primer (5′-GCAAATGGCATTCTGACATCC-3′).

2.5. Analysis of Methanol Utilization Phenotype

1. 10× YNB (13.4% yeast nitrogen base with ammonium sulfate without amino acids): dissolve 134 g YNB with ammonium sulfate and without amino acids in 1 L water, heating if necessary, and filter sterilize. Store at room temperature.

2. 100× H (0.4% histidine): dissolve 400 mg l-histidine in 100 mL water. Heat the solution, if necessary, to no greater than 50°C to dissolve. Filter sterilize and store at 4°C.

3. 10× glucose (20% w/v).

4. 500× B (0.02% biotin): dissolve 20 mg biotin in 100 mL water and filter sterilize. Store at 4°C.

5. MMH (minimal medium containing methanol and histidine) per liter: 1.34% YNB, 4×10^{-5}% biotin, 0.5% methanol, 10 mL 100× H solution, 15 g agar. YNB, methanol, biotin, and histidine are added post-sterilization. For MMH agar, add 15 g agar.

6. MDH (minimal medium containing glucose and histidine) per liter: 1.34% YNB, 4×10^{-5}% biotin, 2% glucose, 10 mL 100× H solution, 15 g agar. YNB, glucose, biotin, and histidine are added post-sterilization. For MDH agar, add 15 g agar.

2.6. Small-Scale Expression Screening

1. BMGY (buffered glycerol-complex medium): 1% yeast extract, 2% peptone, 100 mM potassium phosphate pH 6.0, 1.34% YNB, 4×10^{-5}% biotin, 1% glycerol.

2. BMMY (buffered methanol-complex medium): 1% yeast extract, 2% peptone, 100 mM potassium phosphate pH 6.0, 1.34% YNB, 4×10^{-5}% biotin, 0.5% methanol.

3. 100% HPLC grade methanol.

4. 24-well plates with gas permeable lids (Whatman), microbioreactor plates (Pall Life Sciences) or baffled shake flasks.

5. Appropriate antibiotic (e.g., 100–2,000 μg/mL zeocin).

2.7. Large-Scale Production in a Bioreactor

1. Basal salts medium (batch phase): per liter 40 g glycerol, 0.9 g $CaSO_4$, 14.67 g K_2SO_4, 11.67 g $MgSO_4 \cdot 7 H_2O$, 9 g $(NH_4)_2SO_4$, 166 mL (15%) hexametaphosphate (added post-sterilization), 833 mL distilled water, 4 mL PTM_1 salts (added post-sterilization; see Note 1).

2. PTM_1 trace salts: per liter 6.0 g $CuSO_4 \cdot 5H_2O$, 0.08 g NaI, 3.0 g $MnSO_4 \cdot H_2O$, 0.2 g $Na_2MoO_4 \cdot 2H_2O$, 0.02 g boric acid, 0.5 g $CoCl_2 \cdot 6H_2O$, 20.0 g $ZnCl_2$, 65.0 g $FeSO_4 \cdot 7H_2O$, 0.2 g biotin, 5.0 mL sulfuric acid. Filter sterilize and store at 4°C in the dark. Shelf life for this is 1 month.

3. Fed batch feed (biomass accumulation phase): 50% w/v glycerol containing 12 mL/L PTM_1 salts.

4. Induction phase feed (production phase): 100% methanol or a mixed feed of methanol and sorbitol (e.g., 60% methanol, 40% sorbitol). Supplement with PTM_1 salts (12 mL/L).

5. Sodium hexametaphosphate solution: add 150 g sodium hexametaphosphate $((NaPO_3)_6)$ to 1 L distilled water, filter sterilize and store at room temperature.

2.8. Propidium Iodide Staining

1. Phosphate buffered saline (PBS).

2. 2 mg/mL propidium iodide (PI) in PBS (store wrapped in foil ≤1 month at 4°C).

3. Cell suspension.

4. 13×100 mm polystyrene culture tubes.

2.9. Analysis of Cell Pellet (Small Scale)

1. Breaking buffer: 50 mM sodium phosphate pH 7.4, 1 mM PMSF (phenylmethylsulfonyl fluoride or other protease inhibitor), 1 mM EDTA, 5% glycerol.
2. Sterile acid-washed glass beads (Sigma).
3. FastPrep (FP120; Thermo Electron Corporation).
4. Table-top centrifuge.
5. Ice.

2.10. Sterility Testing

1. Microscope.
2. Microscopic slides.
3. Cover slips.
4. Mineral oil.

2.11. Gas Chromatographic Analysis of Methanol

1. Gas Chromatograph (Unicam 610 GLC Gas Chromatograph with ProGC software). *See* also Chapter 15.

3. Methods

3.1. Plasmid Linearization

An important step in the generation of a recombinant *P. pastoris* strain is the integration of the vector, which harbors the gene of interest, into the host genome. Unlike other expression systems, where a circular plasmid is transformed into the host and maintained episomally, in *P. pastoris* the plasmid is more typically linearized and integrated into the genome by homologous recombination. The steps from linearization to integration into the host are described below.

1. Measure the plasmid DNA concentration in μg/mL.
2. Set up a restriction digest by choosing the appropriate enzyme (for instance *Pme*I, *Sac*I, or *Bst*XI for pPICZA, B or C).
3. Digest 10–50 μg plasmid DNA with the appropriate enzyme according to the manufacturer's instructions.
4. Using agarose gel electrophoresis, compare a small aliquot of the digest with the undigested plasmid to check for complete linearization. Note that complete linearization of the vector is important for generating a stable clone.
5. After complete linearization, heat inactivate the enzyme or add EDTA to stop the reaction.
6. Purify the digest using either phenol:chloroform extraction or by using a commercial clean-up kit. For phenol:chloroform extraction, add 1 volume of phenol chloroform to the digest and extract once, by vortexing and centrifuging at $18,000 \times g$

for 20 min in a bench-top microfuge or until the phases are clearly separated.

7. Remove the upper phase, without touching the interface, into a new 1.5-mL eppendorf.

8. Add 2.5 volumes of 100% ice-cold ethanol and 0.1 volume 3 M sodium acetate and incubate at −20°C for 1 h.

9. Mix and centrifuge at $18,000 \times g$ for 30 min in a bench-top microfuge to pellet the DNA.

10. Remove the supernatant.

11. Wash the pellet with ice-cold 70% ethanol and centrifuge at $18,000 \times g$ for 20–30 min in a bench-top microfuge.

12. Remove the ethanol, air-dry the pellet, and resuspend in 10 μL sterile, deionized water. Use immediately or store at −20°C until further use.

3.2. Preparation of Competent P. pastoris *Cells*

1. Grow a 5 mL YPD culture overnight of the required *P. pastoris* strain (see Table 2) from a single colony. Grow in a 30°C shaking incubator at 220 rpm.

2. The next day, dilute the overnight culture to a final OD_{600} of 0.15–0.20 in 50 mL YPD. Use a baffled flask large enough to provide good aeration, with the culture accounting for no more than 20% of the total volume.

3. Grow the cells to an OD_{600} of 1–1.2 in a 30°C shaking incubator at 220 rpm. This should take 4–8 h.

4. Centrifuge the culture at $2,000 \times g$ for 5 min and pour off the supernatant.

5. Resuspend the pellet in 9 mL of ice-cold BEDS solution.

6. Incubate the cell suspension for 5 min at 100 rpm in a 30°C shaking incubator.

7. Centrifuge the culture again at $2,000 \times g$ for 5 min at room temperature and resuspend the cells in 1 mL (or 0.02 volumes) of BEDS solution without DTT. Divide into 40-μL aliquots and store at −80°C.

8. The competent cells are now ready for use and can be stored for at least 6 months at −80°C.

3.3. Transformation by Electroporation

Transformation of *P. pastoris* with linearized recombinant plasmids can be carried out by electroporation. Include a negative control comprising an empty vector.

1. Thaw competent cells on ice just before use. Do not vortex them.

2. Thaw the linearized and purified plasmid DNA on ice and mix it slowly with the competent cells using a pipette.

3. Incubate the mixture on ice for 5 min. Transfer into a prechilled electroporation cuvette.

4. Place the cuvette into an electroporator and pulse at 1,800 V for 15 ms pulse length.

5. Quickly revive the cells by transferring 500 μL 1 M chilled sorbitol into the cuvette.

6. Resuspend the cells slowly in sorbitol.

7. Add 500 μL YPD medium and mix slowly with a pipette, then transfer into a 15-mL sterile tube.

8. Incubate the tubes with agitation at 220 rpm, 30°C for 2–3 h.

9. After incubation, remove the tubes and centrifuge at $2,000 \times g$ for 5 min.

10. Remove 200 μL supernatant and resuspend the cells in the remaining supernatant.

11. Spread 100–200 μL of the suspension on YPDS plates supplemented with the highest concentration of antibiotic (2,000 μg/mL; see Note 2).

12. Spread 50 μL of the suspension on the YPDS plates supplemented with all other concentrations of antibiotic.

13. Incubate the plates at 30°C for 5–7 days.

14. Pick a total of around 20 colonies from the plates and purify by streaking on fresh YPD or YPDS plates containing the appropriate concentration of antibiotic. This is the "master plate."

3.4. Confirming Genomic Integration of the Plasmid

The following protocol can be used for checking integration by PCR.

1. For each of the colonies from Subheading 3.3, pick a pinpoint-sized portion from the master plate and resuspend the cells in a micro-centrifuge tube containing 5 μL 10× Taq DNA polymerase reaction buffer. Note that taking more cells may inhibit the PCR reaction.

2. For each reaction, prepare the following: 5 μL deoxyribonucleoside triphosphates (each 2.5 mM), 1 μL *AOX1* 5′ primer (20 pmol/μL), 1 μL *AOX1* 3′ primer (20 pmol/μl), 2 U *Taq* DNA polymerase. Add H_2O to a final volume of 45 μL. For 20 reactions multiply the volumes by 21 to allow for pipetting errors. This is the master mix.

3. Aliquot 45 μL of the master mix into each of the 20 tubes containing colony suspension and spin briefly.

4. Put the samples into a thermal cycler and heat at 95°C for 5 min. Cycle 30 times as follows: 95°C, 1 min; 60°C, 1 min; 72°C, 2 min. Add a final 10 min extension step at 72°C.

5. Run 10 μL of the reaction on an agarose gel of an appropriate percentage and check for bands of the correct size.

6. Clones containing the correct insert should be archived in 20% glycerol for long-term storage at –80°C.

3.5. Analysis of Methanol Utilization Phenotype

Strains X33, GS115, and SMD1168 are Mut[+] strains. These strains can use methanol faster than KM71H, which is a Mut[S] strain. Linearizing a plasmid with different enzymes can give rise to different phenotypes as shown in Table 3. His[+] transformants can be screened for their Mut[+] and Mut[S] phenotype, as described below.

1. Using a sterile toothpick, pick a single colony and streak on both an MMH and an MDH plate, making sure to streak the MMH plate first.

2. Repeat for each colony to be tested.

3. Incubate the plates at 30°C for 2–4 days.

4. After 2 days at 30°C, Mut[+] transformants should have grown well on both MDH and MMH plates, whereas Mut[S] transformants will have grown well only on the MDH plates.

5. Based upon the phenotype, clones can be induced with appropriate amounts of methanol in scale up.

3.6. Small-Scale Screening

1. For each colony to be tested, aseptically transfer to 5 mL BMGY medium in a 50-mL tube and culture for 24 h.

2. Aliquot 3 mL BMMY medium into each well of a sterile 24-well plate and cover with its lid.

3. Inoculate BMMY medium with the BMGY culture to a final OD_{600} of 1. Prior to inoculation, centrifuge the required amount of culture to inoculate the BMMY medium and discard the supernatant. This will remove any residual glycerol in the medium.

4. Culture at 30°C, 220 rpm in a shaking incubator.

5. Supplement the cultures with 0.5–1% methanol every 24 h up to 72 h post-induction.

6. Extract a 500 μL sample from the cultures aseptically every 24 h into 1.5-mL Eppendorf tubes and centrifuge at $18,000 \times g$ for 10 min to separate the supernatant and pellet.

7. Store both at –80°C for subsequent analysis.

3.7. Screening Under Low Aeration Conditions

Some proteins may be produced most efficiently under conditions of low aeration (4), as described in this protocol.

1. Grow a single colony in 5 mL BMGY in a 25 mL sterile, baffled shake flask at 30°C, 220 rpm for 24 h.

2. Add BMMY to a sterile non-baffled flask to a final volume of one third of the flask volume.

3. Inoculate the BMMY with the 24 h culture to a final OD_{600} of 1. If necessary remove any residual glycerol by washing the cells in BMMY before inoculation.

4. Cover the shake flask tightly with a foam bung or a fitted lid.

5. Incubate at 30°C at 150–180 rpm to limit aeration (see Note 3).

6. Take samples every 12–24 h until 96 h post-induction (see Notes 4 and 5).

7. Low aeration can also be mimicked under normal conditions (250–300 rpm) by adding supplements such as cobalt chloride (50–100 μM) or deferoxamine (50–100 μM). The optimum concentration should be determined for each production process and added at 0, 24, 48, 72, and 96 h post-induction.

8. Record OD_{600} at each time point.

9. Store both pellet and supernatant at –80°C for analysis.

3.8. Large-Scale Production in Bioreactors

The three stages of a protein production experiment under the control of the *AOX* promoter are the initial batch phase on glycerol, the fed-batch or biomass accumulation phase, and the induction or production phase, with methanol as the inducer (17). The batch phase typically uses 40 g/L glycerol in the basal salts medium giving OD_{600} of 50–70. Sterile PTM_1 salts (4 mL/L) are added aseptically to the bioreactor post-sterilization. At the end of the batch phase (after approximately 24 h), a sharp spike in the dissolved oxygen trace indicates the exhaustion of the carbon source. This is followed by a fed-batch phase to increase the biomass yield before induction by pumping 50% w/v glycerol at 20 mL/L/h for 4 h. Other feed rates (e.g. 10 mL/L/h) have also been used over a longer time period. The fed-batch phase on glycerol also helps in derepression of the *AOX* promoter. Following the fed-batch phase, there is a transition phase of 2–4 h to acclimatize the cells to a change in carbon source prior to induction. During this transition phase, 1 mL 100% methanol can be added per 1 L culture to allow for a smooth transition into the induction phase, which comprises feeding methanol alone or with another carbon source over a time period of 24–96 h. Pre-calibrated pumps are required to achieve the desired feed rates into the bioreactor.

3.8.1. Inducing with 100% Methanol

1. Prepare a sterile feed bottle and add 100% methanol supplemented with PTM_1 salts (12 mL/L).

2. Connect the feed bottle to the bioreactor aseptically through a pre-calibrated feed pump.

3. During the optimization of a new process and in the absence of a methanol probe, take regular offline samples for gas chromatographic analysis to monitor the residual methanol.

3.8.2. Inducing with a Mixture of Methanol and Sorbitol

1. Prepare a sterile feed bottle with tubing.

2. Transfer a solution of 60% sorbitol, 40% methanol into the feed bottle.

3. Add 12 mL/L PTM_1 salts.

4. Connect the feed bottle to a pre-calibrated pump and make connections to the bioreactor aseptically.

3.8.3. Minimizing Protease Activity

Degradation of recombinant proteins is common due to misfolding and aggregation and can lead to truncations. Several heterologous proteins, including single-chain antibodies (scFv), have been successfully produced at temperatures as low as 15°C (18), and more often in the range 23–28°C (18) to alleviate this problem. Lowering the pH to 3 or increasing it to 7 at the point of induction may also reduce the release of proteases (18). Addition of 0.1–5% casamino acids as a supplement may also have the same effect (18). The amount required should be determined experimentally and added either every 24 h post-induction or as a supplement in the induction medium. Colorimetric and fluorescent protease assay kits are available from Sigma-Aldrich® and Bioprobes Ltd. for the detection of proteases released either into the medium or into the periplasm.

3.9. Analytical Procedures

It is essential that the bioprocess is monitored throughout. Offline samples should be taken aseptically during each stage of the process.

3.9.1. Microscopy

1. Take an offline sample in a sterile tube.

2. Aseptically dilute the sample in a 1.5-mL Eppendorf tube (see also step 6 below).

3. Pipette about 10–20 µL onto microscope slide. Overlay with a cover slip avoiding air bubbles.

4. Overlay the cover slip with mineral oil.

5. Observe the slide with a phase contrast microscope at 100× magnification.

6. Cells should be clearly separated, if not prepare another slide with higher dilution.

7. Cells should appear reasonably sized (typically 3–4 µm in diameter). Some will be actively dividing. Look for cells with different morphology such as bacterial rods (bacillus) or spheres (cocci). Their presence is indicative of a contaminated culture.

3.9.2. Sampling

1. Sampling volumes should be kept to a minimum to avoid disturbing the culture in the vessel. The sample collected should be sufficient for product estimations, as well as measurements of OD_{600} and dry cell weight. For 1 L cultures, 5–10 mL can be taken every 12–24 h post-induction. Samples should be taken

throughout the growth phase in the batch phase, with one to two samples during and at the end of the fed-batch stage.

2. If the process is not optimized, sampling every 12 h post-induction will help to monitor productivity.

3.9.3. Optical Density

Optical density can be measured either by online monitoring, using probes in the bioreactors or by spectrophotometers at defined wavelength (typically 590–600 nm) at each time point. Optical densities should be measured immediately post-sampling. See also Chapters 12–14.

1. Switch on the spectrophotometer before preparing the sample. Set the appropriate wavelength, typically 600 nm.

2. A blank sample should be measured and this should be same the liquid medium that is used for growing the cells.

3. Make appropriate dilutions of the sample to bring the optical densities within the operating range of the spectrophotometer. Note that each spectrophotometer has different ranges which will be given in the manual.

4. Vortex the diluted sample and aliquot into a cuvette.

5. Lower the cuvette into the slot for measurement.

6. Note the reading and multiply it by the dilution factor to give the final reading.

3.9.4. Analysing Wet Cell and Dry Cell Weights

1. For each sample to be analysed, place an empty, clean, Pyrex tube, which can withstand high temperatures, in an oven at 100°C for 24–48 h. Ideally, perform each measurement in triplicate for each time point.

2. Remove the tubes from the oven and place in a desiccator for 24 h to completely dry them.

3. Weigh the tubes and note down weights of the empty tubes.

4. Vortex each sample to be analysed.

5. Aliquot 5 mL of the sample into a Pyrex tube.

6. Centrifuge at $2,000 \times g$ for 10 min.

7. Discard the supernatant.

8. Weigh the tube containing the biomass.

9. Note the wet cell weight of the 5 mL sample after subtracting the weight of the tube when empty.

10. Place all tubes in an oven at 100°C for 24–48 h.

11. Remove the tubes from the oven and place in a desiccator for 24 h to completely dry the samples.

12. Weigh the samples and note down the dry cell weights of the 5 mL samples after subtracting the weights of the empty tubes.

3.9.5. Estimation of Cell Viability Using Flow Cytometry

Estimation of the viability of the culture post-induction is useful for correlating productivity per cell. There are several probes for online monitoring of viable cells. Another reliable way to measure the number of viable cells in an offline sample is by flow cytometry, which can stain either live or dead cells. This procedure must be performed on a freshly harvested sample.

1. Add 1 μL 2 mg/mL propidium iodide (2 μg/mL final concentration) to approximately 10^6 washed cells suspended in 1 mL PBS in 13×100 mm polystyrene culture tubes.

2. Incubate for 30 min in the dark on ice.

3. Analyse by flow cytometry with an excitation wavelength of 488 nm and an emission wavelength >550 nm. The signal can be amplified logarithmically to distinguish populations of non-viable from viable cells.

3.9.6. Estimation of Product Yield

Product estimations for secreted proteins are primarily done on the supernatant, once it has been separated from the biomass. The pellet should also be analysed to determine the total yield of the product. Product estimations can be done by polyacrylamide gel electrophoresis (PAGE). If an antibody specific for the product of interest is available, immunoblotting and/or ELISA are other options.

1. Aliquot 1 mL samples into labeled, sterile 1.5-mL Eppendorf tubes.

2. Centrifuge the samples at $18,000 \times g$ for 10 min.

3. Separate the supernatants from the pellets.

4. Freeze the supernatants at −80°C until further use; they should contain the majority of the recombinant protein of interest.

5. Wash the 1 mL pellet with breaking buffer.

6. Resuspend in 500 μL breaking buffer.

7. Add one tenth of the total volume of acid-washed zirconium beads.

8. Place the tubes in the FastPrep instrument.

9. Homogenize for 30 s at setting 5.5.

10. Separate the supernatant (containing intracellular soluble proteins) from the pellet by centrifugation.

11. Place the supernatant on ice.

12. Repeat steps 6–10 on the pellet four times.

13. Freeze both the pellet and the pooled intracellular supernatants.

14. Perform all the above steps at 4°C.

3.9.7. Methanol Analysis by Gas Chromatography

Vectors with an *AOX* promoter require methanol for induction. Concentrations of methanol in the culture are critical for product accumulation and should not exceed 5 g/L. Process development therefore requires maintaining methanol concentrations below this critical concentration. Methanol levels can be estimated by gas chromatography to optimize the feed rates of methanol post-induction.

1. Aliquot 1 mL 100% methanol (HPLC grade) into a sterile 1.5-mL Eppendorf tube.

2. Make serial dilutions of 0.1–1% methanol with sterile distilled water from 100% methanol (the upper standard depends upon the sensitivity of the column).

3. Vortex the samples thoroughly.

4. Prepare a blank of sterile distilled water.

5. Perform 1 μL injections of each sample to construct a standard curve. Use the peak areas from each standard including the blank.

6. Obtain the equation for the standard curve.

7. Prepare three sets of samples: neat, 1:5, and 1:10 dilution from supernatant of the culture. A higher dilution is required if higher methanol rates are used.

8. Vortex the samples and inject as for the standards.

9. Use the equation for the standard curve to calculate the methanol concentrations in the supernatants.

10. Multiply the value by the dilution factor to get the methanol concentrations where appropriate.

4. Notes

1. The basal salts medium recipe given in Subheading 2.7 prevents precipitation after addition of PTM_1 salts and also when the pH is reduced to 3.

2. Plating out the transformation mixture on higher concentrations of antibiotic selects for "jackpot" clones, which contain multiple copies of the gene. This may increase the yield of recombinant protein, although the correlation between gene dosage and protein yield has not been generically established (19).

3. Antifoams are useful to decrease foaming and increase aeration, and may also affect the yield of the recombinant protein (20). Antifoam should not be added to low aeration experiments.

4. Unless otherwise stated all the experiments are done in triplicate under identical conditions.

5. All experiments must be conducted under strictly sterile conditions.

6. Pretreatment of Pichia pastoris with 0.1 M lithium acetate (LiAc) and 10 mM Dithiothrietol (DTT) before electroporation increases transformation efficiency by several folds.

References

1. Cregg JM, Cereghino JL, Shi J, Higgins DR (2000) Recombinant protein expression in *Pichia pastoris*. Mol Biotechnol 16:23–52

2. Bill RM, Winter PC, McHale CM, Hodges VM, Elder GE, Caley J, Flitsch SL, Bicknell R, Lappin TR (1995) Expression and mutagenesis of recombinant human and murine erythropoietins in *Escherichia coli*. Biochim Biophys Acta 1261:35–43

3. Cereghino JL, Cregg JM (2000) Heterologous protein expression in the methylotrophic yeast *Pichia pastoris*. FEMS Microbiol Rev 24:45–66

4. Baumann K, Carnicer M, Dragosits M, Graf AB, Stadlmann J, Jouhten P, Maaheimo H, Gasser B, Albiol J, Mattanovich D, Ferrer P (2010) A multi-level study of recombinant *Pichia pastoris* in different oxygen conditions. BMC Syst Biol 4:1–22

5. Kim SJ, Lee JA, Kim YH, Song BK (2009) Optimization of the functional expression of *Coprinus cinereus* peroxidase in *Pichia pastoris* by varying the host and promoter. J Microbiol Biotechnol 19:966–971

6. Inan M, Meagher MM (2001) Non-repressing carbon sources for alcohol oxidase (AOX1) promoter of *Pichia pastoris*. J Biosci Bioeng 92:585–589

7. Waterham HR, Digan ME, Koutz PJ, Lair SV, Cregg JM (1997) Isolation of the *Pichia pastoris* glyceraldehyde-3-phosphate dehydrogenase gene and regulation and use of its promoter. Gene 186:37–44

8. Resina D, Cos O, Ferrer P, Valero F (2005) Developing high cell density fed-batch cultivation strategies for heterologous protein production in *Pichia pastoris* using the nitrogen source-regulated *FLD1* promoter. Biotechnol Bioeng 91:760–767

9. Mandenius CF, Brundin A (2008) Bioprocess optimization using design-of-experiments methodology. Biotechnol Prog 24:1191–1203

10. Bawa Z, Bland CE, Bonander N, Bora N, Cartwright SP, Clare M, Conner MT, Darby RA, Dilworth MV, Holmes WJ, Jamshad M, Routledge SJ, Gross SR, Bill RM (2011) Understanding the yeast host cell response to recombinant membrane protein production. Biochem Soc Trans 39:719–723

11. Cos O, Ramón R, Montesinos JL, Valero F (2006) Operational strategies, monitoring and control of heterologous protein production in the methylotrophic yeast *Pichia pastoris* under different promoters: a review. Microb Cell Fact 5:1–20

12. Zhang W, Bevins MA, Plantz BA, Smith LA, Meagher MM (2000) Modeling *Pichia pastoris* growth on methanol and optimizing the production of a recombinant protein, the heavy-chain fragment C of botulinum neurotoxin, serotype A. Biotechnol Bioeng 70:1–8

13. Holmes WJ, Darby RA, Wilks MD, Smith R, Bill RM (2009) Developing a scalable model of recombinant protein yield from *Pichia pastoris*: the influence of culture conditions, biomass and induction regime. Microb Cell Fact 8:35

14. Zhang W, Hywood Potter KJ, Plantz BA, Schlegel VL, Smith LA, Meagher MM (2003) *Pichia pastoris* fermentation with mixed-feeds of glycerol and methanol: growth kinetics and production improvement. J Ind Microbiol Biotechnol 30:210–215

15. Hellwig S, Emde F, Raven NPG, Henke M, van der Logt P, Fischer R (2001) Analysis of single-chain antibody production in *Pichia pastoris* using on-line methanol control in fed-batch and mixed-feed fermentations. Biotechnol Bioeng 74:344–352

16. Zhang Y, Liu R, Wu X (2007) The proteolytic systems and heterologous proteins degradation in the methylotrophic yeast *Pichia pastoris*. Ann Microbiol 57:553–560

17. Cereghino GP, Cereghino JL, Ilgen C, Cregg JM (2002) Production of recombinant proteins in fermenter cultures of the yeast *Pichia pastoris*. Curr Opin Biotechnol 13:329–332

18. Shia X, Karkutb T, Chamankhahc M, Alting-Meesb M, Hemmingsenb SM, Hegedus D (2003) Optimal conditions for the expression of a single-chain antibody (scFv) gene in *Pichia pastoris*. Protein Expr Purif 28:321–330

19. Oberg F, Sjöhamn J, Conner MT, Bill RM, Hedfalk K (2011) Improving recombinant eukaryotic membrane protein yields in *Pichia pastoris*: the importance of codon optimization and clone selection. Mol Membr Biol 28:398–411

20. Routledge SJ, Hewitt CJ, Bora N, Bill RM (2011) Antifoam addition to shake flask cultures of recombinant *Pichia pastoris* increases yield. Microb Cell Fact 10:17

Chapter 20

Disruption of Yeast Cells to Isolate Recombinant Proteins

Mohammed Jamshad and Richard A.J. Darby

Abstract

Yeast is a proven host for the production of recombinant proteins, which may be incorporated in cellular membranes or localized in subcellular compartments. In order to gain access to these proteins, cellular disruption is required to permit extraction, purification, and downstream analysis. Disruption can significantly impact the yield and quality of the biomaterial. We highlight several disruption techniques that are applicable to yeast cells ranging from mechanical to nonmechanical approaches. In all cases fast, efficient cellular disruption is desirable, that does not alter the protein chemically or physically and that generates material for downstream purification and analysis.

Key words: Homogenization, Disruption, French press, Emulsiflex C3, FastPrep

1. Introduction

After successfully identifying recombinant protein-producing clones and optimizing their scale-up, the cells must be collected (usually from bioreactors) by filtration or centrifugation. Cellular disruption and isolation of membrane and/or soluble fractions can then be conducted either immediately or the wet cell mass can be snap frozen in liquid nitrogen (in thin layers for fast cooling) and stored at −80°C until required. Depending both on the growth conditions and growth phase in which the cells are harvested different disruption approaches may be required. Consequently, the cells should be viewed under a microscope after disruption to determine the efficiency of the breakage.

Mechanical or nonmechanical disruption can be applied to yeast cells. Mechanical disruption systems include bead mills, homogenizers, and jet streams (1, 2). Nonmechanical disruption methods can be chemical and enzymatic (lyticase or zymolase (3–5)) or physical (decompression, sonication, thermolysis, and osmotic

Roslyn M. Bill (ed.), *Recombinant Protein Production in Yeast: Methods and Protocols*, Methods in Molecular Biology, vol. 866, DOI 10.1007/978-1-61779-770-5_20, © Springer Science+Business Media, LLC 2012

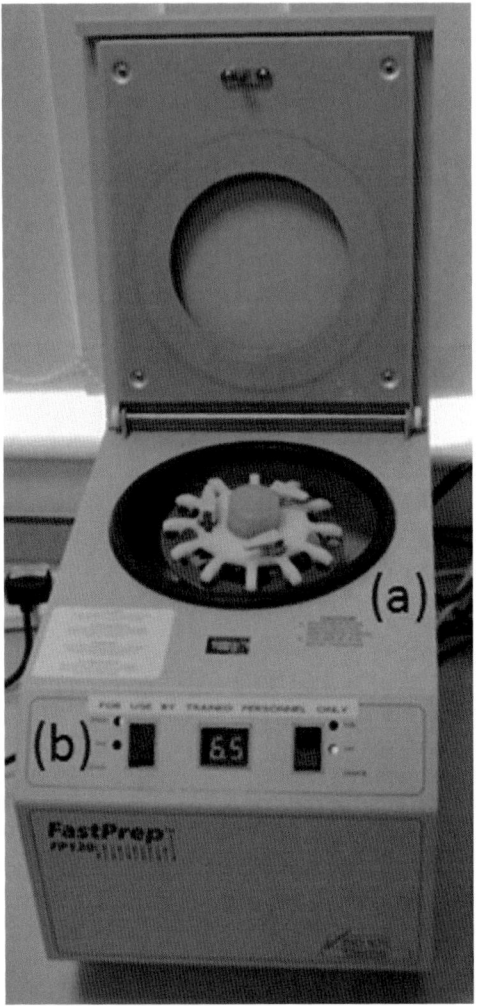

Fig. 1. A photograph showing the Thermo FastPrep FP120 in the laboratory of Dr Roslyn Bill, Aston University, UK. The 2-mL sample tubes are loaded into the rotor (**a**) and the locking ring (shown in the closed position) is used to secure them in position. The lid is then closed, and the operating conditions, as outlined in Subheading 3.1, are set using the control panel (**b**). Upon completion of the cycle, the locking ring is opened and the samples are placed on ice for 2 min to dissipate any latent heat.

shock (6–8)). Some of these methods are more appropriate for transformation and subcellular fractionation rather than the general disruption of yeast cells. The authors' preferred method is mechanical disruption, although we present one chemical method for those laboratories without access to the equipment required for mechanical breakage. However, it should be noted that this method is suitable only for the extraction of soluble proteins.

For the disruption of small samples (less than 2 mL), acid-washed glass beads are very effective (see Subheading 3.1) when using a FastPrep (FP120; Thermo Electron Corporation) (9) (Fig. 1).

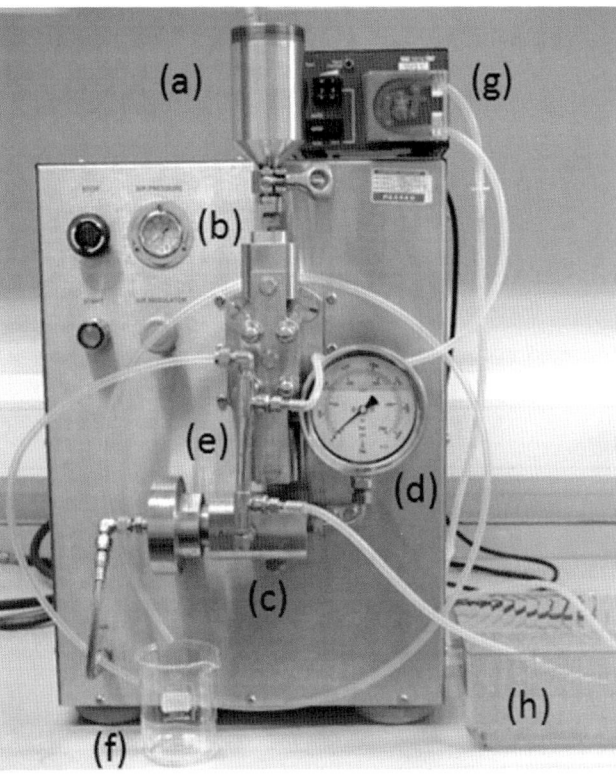

Fig. 2. A photograph showing the experimental set up of the Avestin EmulsiFlex-C3 in the laboratory of Dr Roslyn Bill, Aston University, UK. The equipment is first switched on and the start-up procedure carried out as outlined in Note 4. The sample is then poured into the reservoir (**a**) and the air pressure set to 40 psi, as determined by the gauge (**b**). Once the homogenization value (**c**) starts compressing, the air pressure should then be increased until the homogenization pressure reaches 30,000 psi, as determined by the gauge (**d**). The sample will then pass through the chilled heat exchanger (**e**) and into the collection reservoir (**f**) via the outlet tube, or be returned to the sample reservoir for repeated processing. The heat exchanger can be chilled by connecting to an external peristaltic pump (**g**) and circulating ice-chilled water (**h**). It should be noted that the collection reservoir is not shown on ice for clarity.

The disadvantage of this method is that latent heat is produced during the mechanical disruption and therefore the samples must be cooled on ice between runs. If a FastPrep or equivalent equipment is not available, it is possible to use a vortexer and to process the samples by hand; however, the efficacy of disruption is significantly reduced (see Subheading 3.2).

Cultivation of yeast in a bioreactor generates large biomass yields for which constant cell disruption systems are very convenient and efficient. The EmulsiFlex-C3 marketed by Avestin (Fig. 2) has a constant flow rate of 3 L/h, adjustable homogenizing pressure between 500 and 30,000 psi (7 and 2,000 bar) and a heat exchanger, enabling up to 200 mL of suspended cells to be continuously processed (2, 9). When using this method to disrupt the cells, it may be

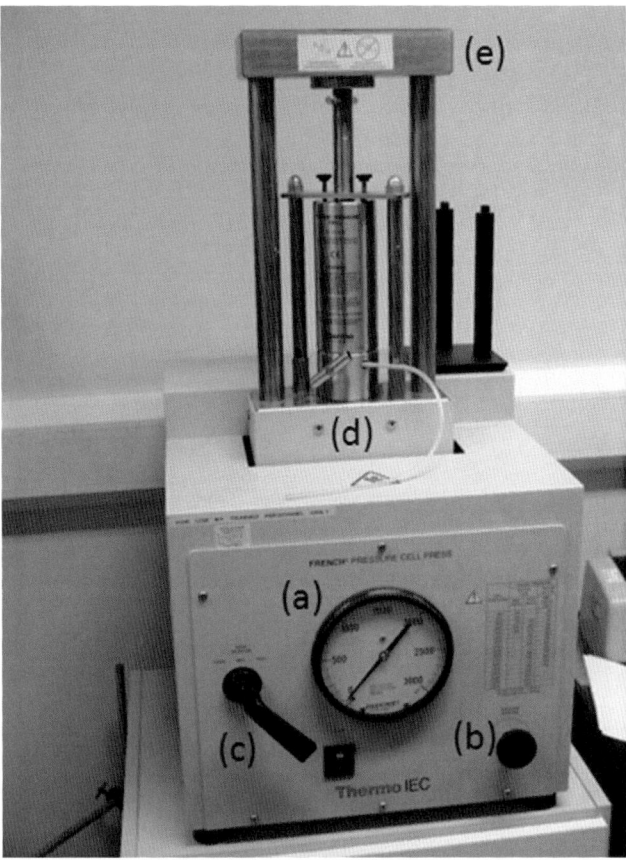

Fig. 3. A French Press with the 40 K cell in the operational position within the laboratory of Dr. Roslyn Bill, Aston University, UK: (a) pressure gauge, (b) pressure increase knob, (c) ratio selector handle, (d) hydraulic platform, (e) platen.

necessary to process the sample more than once (as determined by observing the cells under a microscope and assessing the percentage disrupted—the authors recommend a minimum of three passages).

The French Press (ThermoFisher) is a motor-driven press that can disrupt yeast cells using variable pressure, depending on the model of cell used (Figs. 3 and 4). As the pressure increases in the homogenization cell, the intracellular pressure also increases. As the sample is dispensed from the outlet tube, the external pressure on the yeast cell walls decreases rapidly to atmospheric pressure. The intracellular pressure also decreases but more slowly and it is this pressure differential that ruptures the cell wall membranes. The standard cell has a maximum working pressure of 40,000 psi and capacity of 35 mL. For smaller sample volumes, there is a 20,000 psi cell, with a capacity of 1.4–3.7 mL. The method described below is based on the larger 35 mL cell.

Irrespective of the method of breakage, the separation of cellular debris, soluble proteins, and membrane particles can be achieved

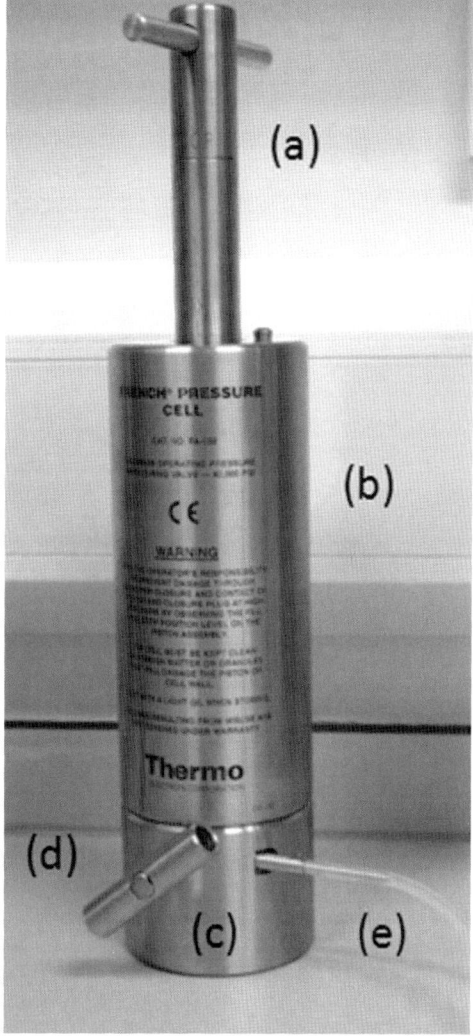

Fig. 4. Close-up of the assembled 40 K cell: (**a**) piston, (**b**) 40 K cell body, (**c**) 40 K cell lid, (**d**) outlet valve, (**e**) sample outlet and tube.

in two steps: low-speed centrifugation to spin down cell debris followed by high-speed centrifugation of the supernatant. Further isolation of the plasma membrane fraction can be achieved if required using standard sucrose gradient methods (10).

2. Materials

1. Acid washed glass beads: 0.3–0.5 µm diameter.
2. Phenylmethylsulfonyl fluoride (PMSF): *store at 4°C.*
3. Yeast-specific protease inhibitor cocktail (Merck): *store at −20°C.*

4. 1.5–2.0 mL screw-cap tube.

5. FastPrep (FP-120), for small-scale preparations (Thermo Electron Corporation).

6. EmulsiFlex-C3, for large-scale preparations (Avestin).

7. French Press, for large-scale preparations (ThermoFisher).

8. Breaking Buffer: 50 mM sodium phosphate pH 7.4, 5% glycerol, 2 mM EDTA, 100 mM NaCl, *store at 4°C.*

9. Buffer A: 20 mM HEPES pH 7.4, 50 mM NaCl, 10% glycerol, *store at 4°C.*

10. Homogenizer; PTFE Pestle/S.S Rod (Scientific Laboratory Supplies).

11. YeastBuster: *stable at room temperature* (Merck).

12. Benzonase: *store at -20* (Merck).

3. Methods

Unless it is otherwise stated, all the following methods should be carried out on ice.

3.1. Acid-Washed Glass Beads (<0.5 g Cells)

1. Mix 100–300 mg of wet yeast cells with 300–500 μL of acid-washed glass beads and 300–500 μL of ice-cold Breaking Buffer (see Note 1) in a 2-mL screw-cap tube. Add PMSF to a final concentration of 2 mM (other protease inhibitors may be added as well, according to the manufacturer's instructions).

2. Agitate the sample in a FastPrep (FP-120) for a period of 45 s at 6,500 rpm. Place the sample immediately on ice for 2 min to dissipate any residual heat.

3. Repeat step 2 at least five more times.

4. Following disruption, observe the cells under a light microscope to ascertain the extent of cell breakage (>80% should be disrupted). Repeat steps 2 and 3 if necessary.

5. To separate the supernatant and cell debris from the glass beads, pierce the bottom of the tube with a flamed hypodermic needle and place the tube into the cap of a modified 15-mL Falcon tube (see Note 2). Recover the supernatant by centrifugation at $5,000 \times g$ at 4°C for a period of 3 min.

6. Clarify the supernatant at $10,000 \times g$ for 30 min at 4°C and recover the supernatant. This is a total cell lysate containing both soluble and membrane proteins.

7. To isolate membranes from the total cell lysate, centrifuge the sample at $100,000 \times g$ for 60 min at 4°C and resuspend the pellet in ice-cold Buffer A.

3.2. Acid-Washed Glass Beads (5–10 g Cells)

1. Resuspend 5 g of wet cells in 10 mL of ice-cold Breaking Buffer and PMSF to a final concentration of 2 mM (other protease inhibitors may be added as well, according to the manufacturer's instructions). Add an equivalent pellet volume of acid-washed glass beads, vortex the suspension for 5 min, and subsequently cool on ice for a period of 5 min.

2. Repeat the vortex and cooling steps for 15 cycles.

3. Determine the extent of cell breakage (as outlined in Subheading 3.1) and repeat the disruption if necessary.

4. Clarify the supernatant at $10,000 \times g$ for 30 min at 4°C and recover the supernatant. This is a total cell lysate containing both soluble and membrane proteins.

5. To isolate membranes from the total cell lysate, centrifuge the sample at $100,000 \times g$ for 60 min at 4°C and resuspend the pellet in ice-cold Buffer A.

3.3. Disruption by Chemical Lysis

1. Centrifuge the culture at $3,000 \times g$ for 10 min at 4°C and discard the supernatant.

2. Determine the wet weight of the pellet and resuspend it in YeastBuster Reagent (5 mL) and THP Solution (50 μL) per gram of wet cell. This step is done at room temperature.

3. Add 25 units of Benzonase/mL of YeastBuster plus protease inhibitors as directed by the manufacturers.

4. Incubate the sample at room temperature on a platform rocker or rotor mixer at the lowest setting for 30 min.

5. Centrifuge the sample at $16,000 \times g$ for 20 min at 4°C and store the recovered supernatant at –20°C if it is not required immediately (see Note 3).

3.4. High-Pressure EmulsiFlex-C3 (>10 g Cells)

1. Resuspend the cell pellet in ice-cold Breaking Buffer at a ratio of 3:1 buffer (mL) to pellet (g). Ensure an even resuspension of the cells using a handheld homogenizer and pestle.

2. Carry out the start-up/shut-down procedure (see Note 4).

3. Pass the cell suspension through the Emulsiflex-C3 cell disrupter fitted with a chilled heat exchanger (Avestin) three times according to the manufacturer's instructions (see Note 5).

4. Observe the cells under a light microscope to check the extent of cell breakage. The breaking efficiency should be >90% at a homogenizing pressure of 30,000 psi.

5. Centrifuge the sample at $10,000 \times g$ for 30 min at 4°C to remove any unbroken cells and cellular debris.

6. To isolate the total membrane fraction from the clarified lysate, centrifuge the sample at $100,000 \times g$ for 60 min at 4°C and discard the supernatant.

7. Resuspend the membrane pellet in ice-cold Buffer A using a glass homogenizer at a ratio of 5–10 mL buffer/g pellet.

3.5. French Press

1. Cool the French Press cell prior to use (overnight in the cold room). Resuspend the cell pellet in ice-cold Breaking Buffer ensuring an even resuspension of the cells in no more than 30 mL if using the 40 K cell (see Notes 6 and 7).

2. Wet the rubber seals of the plunger and lid with glycerol (see Note 8).

3. Place the cell onto the supplied stand and slide in the plunger up to the maximum fill line and invert. Fit collection tubing to the outlet valve and screw into the cell. Leave the valve partially open.

4. Pour the cell suspension into the French Press cell and push down the lid. Excess liquid will come out of the open outlet valve tubing. Close the outlet valve tight, turn the cell over (so the plunger is now at the top) and place in the French Press (see Note 9). Ensure that the cell is properly centered by resting it against the three pins located on the platform with the outlet valve facing the front.

5. Pull the locking arm across the cell and finger-tighten the two screws to lock it in place. This prevents the cell from moving during operation. Raise the piston in the cell and position it so that it contacts the platen (Fig. 3) with its arms fitted in place in the bracket.

6. Switch the lever to DOWN and switch the power ON. Run the platform to the lowest position (see Note 10). Adjust the pressure by turning the pressure increase knob clockwise until the pressure gauge indicates *1,900* (see Note 11).

7. Turn the lever to MEDIUM and wait for the cell to pressurize. The platform will move up until the pressure gauge reads *1,900*. Switch the lever to HIGH. When the pressure gauge reads *1,900*, slowly open the outlet valve to release the cell lysate into a prechilled beaker stored on ice. To ensure efficient breakage, the sample should be released as continuous droplets so that the pressure does not fall significantly. If the valve is open too much then the entire sample will flow out very quickly without efficient disruption taking place.

8. When the entire sample has been expelled, leave the outlet valve open and turn the lever down to MEDIUM and then DOWN in one smooth motion. Wait for the platform to return to the lowest position and remove the cell. Repeat steps 3–6 five times. After the final passage, turn the lever to the DOWN position and remove the cell ready for cleaning.

9. Remove the cell and fill with water. Force the water out through the cell body using the press, following steps 3–6. Repeat three

times and then turn the lever to the DOWN position and dial down the pressure to 0. Remove the cell and rinse with water. Dry with soft lint-free towels (see Note 12).

10. Return the cell to the refrigerator or cold room until next use.

4. Notes

1. This buffer is recommended by the authors; however, alternative buffers that meet the reader's downstream needs can be substituted in all of the disruption methods outlined without adversely affecting the efficiency of disruption.

2. Using a sharp scalpel cut a hole in the middle of the 15-mL tube cap. This hole should be large enough to accommodate the lanced Fastprep tube without letting it fall into the Falcon tube. Upon centrifugation the lysed sample will collect in the 15-mL tube while the glass beads are retained in the Fastprep tube.

3. YeastBuster is not compatible with detergent-sensitive protein assays and it also prevents proteins binding to ion exchange resins. Therefore, it is recommended that extracts are dialyzed prior to analysis or chromatography.

4. *Start-up procedure.* Regulate the incoming air pressure to 100 psi for control of the pneumatic control cylinder. Prior to a sample run, put 400 mL of deionized water through the system under operating conditions to remove residual ethanol left in the system from the shut-down procedure. *Shut-down procedure.* After cell breakage run 200 mL of deionized water through the system under operating conditions. Then prime the pump with 200 mL 20% ethanol, and as soon as the reservoir has emptied switch off the system leaving residual ethanol within the unit.

5. For the disruption of yeast with the High-Pressure Emulsiflex-C3 it is highly recommended that a ceramic stem and seat are fitted to the system. A further recommendation is that the 1 lb outlet check valve spring is replaced with the 14 lb rated spring. This facilitates constant flow of the samples through the homogenization valve and ensures adequate breakage of the cells.

6. Cells should be resuspended at a buffer ratio of 3:1 buffer to pellet (v/w) which ensures the suspension is not too thick.

7. The French Press is extremely heavy; handle with extreme care.

8. Examine the O-ring on the piston *before and after* each use. If it looks worn (it is nicked or black deposits come off when it is touched), replace the O-ring immediately. Breakage of the O-ring during use can result in the piston becoming jammed.

9. Do not overtighten the outlet valve as this can result in damage to the valve or system leaks. It should be finger-tight only.

10. Always check that the pressure is set properly when the platen is in the lowest position (DOWN). Adjust the pressure with the pressure increase knob to the desired value. *DO NOT* adjust the pressure when the cell itself is under pressure.

11. A pressure setting of 1,900 displayed on the gauge equates to approximately 30,000 psi within the cell on the high setting. *DO NOT* exceed a gauge setting of 2,500 when using the 40 K cell.

12. Regular soft paper towels can be used for cleaning and drying the inside of the cell and piston. Hard-bristled brushes should not be used on any part of the French Press.

References

1. Van Gaver D, Huyghebaert A (1991) Optimization of yeast cell disruption with a newly designed bead mill. Enzyme Microb Technol 13:665–671

2. Jamshad M, Rajesh S, Stamataki Z, McKeating JA, Dafforn T, Overduin M, Bill RM (2008) Structural characterization of recombinant human CD81 produced in *Pichia pastoris*. Protein Expr Purif 57:206–216

3. Elorza MV, Rodriguez L, Villanueva JR, Sentandreu R (1978) Regulation of acid-phosphatase synthesis in *Saccharomyces cerevisiae*. Biochim Biophys Acta 521:342–351

4. Knorr D, Shetty KJ, Kinsella JE (1979) Enzymatic lysis of yeast-cell walls. Biotechnol Bioeng 21:2011–2021

5. Scott JH, Schekman R (1980) Lyticase – endoglucanase and protease activities that act together in yeast-cell lysis. J Bacteriol 142:414–423

6. Alvarez P, Sampedro M, Molina M, Nombela C (1994) A new system for the release of heterologous proteins from yeast based on mutant strains deficient in cell integrity. J Biotechnol 38:81–88

7. Magnusson KE, Edebo L (1976) Influence of cell concentration, temperature, and press performance on flow characteristics and disintegration in freeze-pressing of *Saccharomyces cerevisiae* with X-press. Biotechnol Bioeng 18:865–883

8. Blechl AE, Thrasher KS, Vensel WH, Greene FC (1992) Purification and characterization of wheat alpha-gliadin synthesized in the yeast *Saccharomyces cerevisiae*. Gene 116:119–127

9. Bonander N, Darby RAJ, Grgic L, Bora N, Wen J, Brogna S, Poyner DR, O'Neill MAA, Bill RM (2009) Altering the ribosomal subunit ratio in yeast maximizes recombinant protein yield. Microb Cell Fact 8:10

10. Walworth NC, Goud B, Ruohola H, Novick PJ (1989) Fractionation of yeast organelles. Methods Cell Biol 31:335–354

INDEX

Roslyn M. Bill (ed.), *Recombinant Protein Production in Yeast: Methods and Protocols*, Methods in Molecular Biology,
vol. 866, DOI 10.1007/978-1-61779-770-5, © Springer Science+Business Media, LLC 2012

Printed in Great Britain
by Amazon